THE
ANGELS

THE ANGELS

Explanation of the work *Tractatus de Substantiis Separatis* by Saint Thomas Aquinas

Miguel Grosso

First Edition. October 2024
Copyright © 2024 Miguel Alberto Grosso
ISBN: 9798342270892
grossomiguel2005@yahoo.com.ar
Independent Publication
All rights reserved

Original title: *Los ángeles. Explicación de la obra "Tractatus de Substantiis Separatis" de Santo Tomás de Aquino.* (2024)
Author: Miguel Grosso

INDEX

THE PHILOSOPHICAL ROOTS OF TOMISTIC ANGELOLOGY 1
PROLOGUE..5
1. THE OPINION OF THE FIRST PHILOSOPHERS AND PLATO 6
2. ARISTOTLE'S OPINION 19
3. HARMONY AND DISCORD: ARISTOTLE AND PLATO ON IMMATERIAL SUBSTANCES (1) 31
4. HARMONY AND DISCORD: ARISTOTLE AND PLATO ON IMMATERIAL SUBSTANCES (2) 39
5. MATTER AND FORM IN SPIRITUAL SUBSTANCES: THE CASE OF AVICEBRON..47
6. THE MATERIALIST FALLACY OF AVICEBRON............................51
7. BEYOND MATTER: THE UNIQUE NATURE OF SPIRITUAL SUBSTANCES..55
8. RESPONSE TO THE ARGUMENTS OF AVICEBRON......................60
9. THE FIRST CAUSE AND SPIRITUAL SUBSTANCES: A DEFENSE OF CREATION..64
10. GOD AS THE FIRST CAUSE OF ALL SUBSTANCES 72
11. THE UNITY OF PERFECTION IN SPIRITUAL SUBSTANCES: A CRITIQUE OF THE NEOPLATONISTS 78
12. THE INITIAL INEQUALITY OF SPIRITUAL SUBSTANCES: A CRITIQUE OF ORIGEN 83
13. KNOWLEDGE AND PROVIDENCE OF SPIRITUAL SUBSTANCES 88
14. KNOWLEDGE AND PROVIDENCE OF GOD................................94
15. RESPONSE TO OBJECTIONS ABOUT THE KNOWLEDGE AND PROVIDENCE OF GOD 106
16. ERRORS OF THE MANICHEANS 113
17. ORIGIN OF SPIRITUAL SUBSTANCES ACCORDING TO THE CATHOLIC FAITH 118
18. THE IMMATERIAL NATURE OF ANGELS..................................122
19. THE ORIGIN OF EVIL IN ANGELS: A THEOLOGICAL DEBATE 127
BY WAY OF AN EPILOGUE..134
ENDNOTES

THE PHILOSOPHICAL ROOTS OF TOMISTIC ANGELOLOGY[1]

The *Tractatus de Substantiis Separatis* is a work by Saint Thomas Aquinas written for his friend and collaborator Fray Reynaldo de Piperno, who also edited the incomplete works of the Angelic Doctor after his death in 1274. The authenticity of the treatise has never been questioned.

It is noted as a primarily philosophical work. However, it is important to highlight that it remains unfinished, and consequently, the theological part is not fully developed, which leaves the philosophical section as predominant in the work.

(...) it holds special importance for the study of rational psychology. Saint Thomas reviews the various theories that ancient philosophers formulated regarding the nature of angels and their operation. Ultimately, he presents the dogma of faith concerning their nature and operation.[2]

The work is also known by other names, according to the manuscripts found: *Tractatus de Angelis, De Natura Angelorum, Libellus de Angelis sive de Substantiis Separatis,* and *Liber de Angelis.*

It has repeatedly been classified as an *opusculum*. Joseph Eschmann points out that *the notion of opusculum, meaning a smaller or lesser work, lacks a precise meaning in itself and has indeed harmed the cause of Thomistic bibliography, if not Thomistic studies in general.*[3]

While it is shorter than some of the other works called major by the Angelic Doctor, Étienne Gilson qualifies it as *an incomparably rich historical work*, and Eschmann calls it *one of the most important metaphysical writings of Aquinas*. Finally, Karl Henle has described the first chapter as *the most elaborate synthesis of Platonic doctrine found throughout the Thomistic corpus.*[4]

The exact date of its composition is unknown. Since the treatise is unfinished, it is presumed that the work was composed shortly before the death of the Angelic Doctor. This is likely the reason why Étienne Mandonnet (1879-1949) and Louis Glorieux (1908-1991) suggested placing its composition in the years 1272-1273. Louis Walz (1890-1967) proposes a longer period, namely 1261-1269. Hugh Callus (1900-1980) dates the work to 1272. On the other hand, Martin Grabmann (1875-1949) uses May 18, 1268, as the *terminus a quo*, that is, the earliest possible point in time when the book could have been written or completed. This is the date or period marking the beginning of the interval in which the book is estimated to have been drafted, based on historical or textual evidence.[5]

Conversely, the year 1270 has also been proposed as another possible *terminus a quo*. However, Jean Vansteenkiste (1909-2003) disagrees with this date and leans towards locating it in an earlier period, around 1259.[6]

Saint Thomas reveals himself (...) not only as a great philosopher in the constructive sense of a system but especially as an expositor of other systems, so that by investigating the origins of errors, he appears as a great psychologist of the minds of philosophers.[7]

The work is divided into two large parts, though disproportionate:

First Part: Philosophical Examination

This part comprises chapters I-XVII and focuses on an analysis of the opinions of various philosophers regarding separated substances (such as angels), with a detailed critique of each position from a strictly philosophical point of view. Here, Saint Thomas reviews and refutes ideas that do not align with Christian doctrine.

> **Chapter I**: Discusses the opinions of the ancient naturalist philosophers and Plato.

> **Chapter II**: Analyzes the position of Aristotle, who studied separated substances based on motion.
>
> **Chapters III and IV**: Compare the similarities and differences between Plato and Aristotle.
>
> **Chapters V-VIII**: Explain and refute the position of Avicebron, who held that angels are composed of matter and form.
>
> **Chapter IX**: Rejects the opinion of those who denied that separated substances were created.
>
> **Chapter X**: Examines and refutes Avicenna's teaching on the procession of all things from a First Principle.
>
> **Chapter XI**: Critiques the position of the Neoplatonists regarding an order of causality among spiritual substances.
>
> **Chapter XII**: Addresses the error of Origen, who claimed that all spiritual substances were created equal.
>
> **Chapters XIII-XVI**: Study and reject the opinions denying God and the angels knowledge of singulars, as well as divine Providence over human acts.
>
> **Chapter XVII**: Refutes the error of the Manicheans.

Second Part: Presentation of Catholic Teaching

This second part, much shorter, covers chapters XVIII to XX. It presents the teaching of the Catholic faith regarding spiritual substances, based on Sacred Scripture, the Church Fathers, and especially on Pseudo-Dionysius the Areopagite. This part is incomplete, as Saint Thomas did not finish the work.

> **Chapter XVIII**: Expounds what must be upheld according to the Catholic faith about the origin of angels, asserting that all spiritual substances were created directly by God.
>
> **Chapter XIX**: Discusses the nature of angels, noting that they are incorporeal, immaterial, and not united to bodies.
>
> **Chapter XX**: Saint Thomas begins to explain the distinction among spiritual substances, but the chapter and the work remain incomplete.

Due to this, the theological part is not fully developed, which leaves the philosophical section as predominant in the work.

The immense legacy of Saint Thomas Aquinas and his work has partly obscured the fact that in the 13th and 14th centuries he was strongly criticized. The thesis of the angel-species was condemned in 1277 on three occasions. The Franciscan school (Duns Scotus, Peter of Jean Olivi, Peter Auriol, William of La Mare) questions several claims of Thomas; but perhaps Francisco de la Marche distinguishes himself most radically from the positions of the Dominican (Tiziana Suarez-Nani, La matière et l'esprit. Études sur François de la Marche, Freiburg/Paris, Le Cerf, 2015). All these works, both those of Suarez-Nani and the translation by N. Blanc, revive the question of the angel, which had become marginal over time, despite its reactivation by K. Gödel's incompleteness theorem, who indeed believed in angels and in a spiritual hierarchy filled with intelligible essences.[8]

For the purposes of reading the *Tractatus* and its explanation, we will use the text translated from Latin to Spanish titled *De substantiis separatis*, edited by the *Biblioteca de Autores Cristianos (BAC)*, Madrid, 2007.

PROLOGUE

The very brief prologue of Saint Thomas Aquinas to his work *Tractatus de Substantiis Separatis* establishes a context of devotion and study.

*In the prologue of his text, Thomas emphasizes the need to divide his research on angels into two stages: the first dedicated to Antiquity, its conjectures, and what can be used to aid in the understanding of the Christian mystery, and the second showing what the Church's teaching is on the matter.*9

He begins by acknowledging the importance of prayer but justifies his written approach as a way to complement devotional practice.

Since we cannot attend the solemn liturgy in honor of the Angels, we should not let the time for prayer go to waste; therefore, we will compensate with the work of writing what we take away from the office of psalmody.

This intellectual and spiritual approach suggests that the pursuit of knowledge about angels is also a form of prayer.

Next, in the first chapter, he will survey the various conceptions of reality that prevailed in ancient philosophy, from the pre-Socratic philosophers (Thales, Anaximenes, Heraclitus, Empedocles, Anaxagoras, Democritus) to Plato.

What we aim to do is document the excellence of the holy angels with all possible resources; for this purpose, it seems we must first investigate what human conjecture thought of them in ancient times, so that we can embrace whatever we find in accordance with the faith and refute what we see as incompatible with Catholic doctrine.

1. THE OPINION OF THE FIRST PHILOSOPHERS AND PLATO

In Chapter I, Saint Thomas examines the opinions of the ancient philosophers regarding the principles of reality.

These thinkers were called "naturalist philosophers." They were thinkers from ancient Greece who attempted to explain the origin, structure, and principles of the universe based on natural elements and processes, without resorting to mythological or supernatural explanations. These philosophers are considered forerunners of scientific thought, as they sought to understand reality through observations of the physical world and rational theories. They are called "naturalists" because they focused on *physis* (nature), that is, on natural phenomena and the material principles that govern the universe.

Thales of Miletus, Heraclitus, and Empedocles attempted to understand the fundamental nature of reality and the principles that underlie it.

-Thales of Miletus (approx. 624-546 BC): He considered that the basic and fundamental principle of all things was water. According to Thales, water was the primordial element from which everything originates and to which everything returns. He believed that water had the ability to transform into other elements, thus representing the foundation of all existence.

-Heraclitus (approx. 535-475 BC): He thought that the fundamental principle was fire. He argued that change and transformation were essential characteristics of reality, and fire symbolized this constant process of change. According to him, everything is in a state of continuous flux, and fire represents the dynamism and perpetual transformation.

-Empedocles (approx. 495-435 BC): He introduced the idea that the fundamental principles were several elements: water, air, fire, and earth. According to him, these four elements were the roots or basic principles of everything that exists. The elements combine and separate through two opposing forces he called Love and Strife, thereby generating the variety and complexity of the world.

These naturalist philosophers shared the idea that the fundamental or first principles of reality were physical or corporeal elements, in contrast to later ideas that would introduce more abstract or incorporeal principles to explain the nature of the universe.

Democritus and Anaxagoras introduced new ideas regarding fundamental principles.

-Democritus (approx. 460-370 BC): He believed that everything in the universe was composed of atoms, tiny indivisible particles that combine to form all matter. For him, atoms were the material and universal basis of all things.

-Anaxagoras (approx. 500-428 BC): He introduced the idea that, in addition to material parts, there is an incorporeal principle called *nous* (mind or understanding), which organizes and gives shape to things. Although his material principles were minimal parts similar to atoms, he added that there was a separate, non-material principle that distinguished and ordered matter.

Anaxagoras did not recognize the mind as a universal principle of being, but only as a distinctive principle. Furthermore, he did not incorporate incorporeal substances, such as angels, into his reflection.

Anaxagoras introduced the concept of *nous* (understanding or mind) as a principle that organizes and orders mixed matter. According to his theory, the *nous* is not the ultimate cause or essence of being, but a force that separates and organizes the parts of the cosmos. In other words, *nous* is the principle that causes distinction and order but is not the universal or fundamental principle of being.

Therefore, by describing *nous* as a distinctive principle, it is implied that its function is more to separate and organize things rather than to be the ultimate foundation of reality. **This principle does not address the existence of incorporeal substances or the ultimate essence of things**, as more abstract or universal notions of principles do in other philosophies.

And since it was in everyone's mind that the first principle of things should be considered as God, insofar as any of them attributed the category of first principle to a body, they also granted it the name and dignity of the divine. We say all this because **such philosophers and their followers never thought that there existed incorporeal substances, which we call angels**.

Indeed, these philosophers attributed divine characteristics to the bodies considered to be first principles. However, they did not contemplate the existence of incorporeal substances such as angels. In the context of ancient philosophy, the expression "first principles or fundamental principles" refers to the most basic or foundational aspects of reality from which everything else is derived. The "first principles" are the basic elements that explain the existence and nature of the universe.

These ancient philosophers viewed the material bodies they considered first principles as if they had divine qualities. This means:

1-The first principles were not just ordinary material elements. Instead, they were considered entities with fundamental importance in the structure of the cosmos.

2-Attributing divine characteristics implies that these bodies were seen as possessing exceptional or sacred qualities. In this context, this could mean that the philosophers thought that these bodies were not only the basis of reality but also possessed some form of perfection, power, or influence that made them special compared to other material bodies.

The Epicureans, a philosophical school influenced by Democritus, also conceived of gods as corporeal, but unlike the ideas of the earlier philosophers mentioned, these gods did not engage in activities or intervene in the world. For the Epicureans, the gods existed in a state of complete detachment and passivity.

This vision of corporeal and passive gods also resonated with **the Sadducees**, a group within ancient Judaism. The Sadducees were known for their focus on a literal interpretation of the Torah, and like the Epicureans, they did not accept the idea of angels or incorporeal beings.

(...) Plato was the one who adopted a more effective way to overcome the opinions of the first naturalist philosophers.

We say all this because such philosophers and their followers never thought that there were incorporeal substances which we call angels.

Plato's thought (427 B.C. – 347 B.C.) surpasses the limitations of the naturalist philosophers by proposing that the senses cannot provide true knowledge due to the constant change of material bodies. Instead of relying on the material, Plato introduces the idea that there are entities separate from matter that represent fixed truths, enabling the soul to access genuine knowledge.

The naturalist philosophers held that humans cannot reach any certain truth about things, either due to the constant flow of corporeal things or the deceptions caused by the senses with which we perceive bodies. For this reason, Plato proposed certain natures that, being separated from the matter of fluctuating things, would be subjects of fixed truth, such that by adhering to them, the soul could know the truth. Thus, since the intellect, when knowing the truth, apprehends something separate from the matter of sensible things, he maintained that there exist certain natures separated from the sensible.

This passage describes how, due to the difficulty in attaining knowledge of corporeal things due to their constant change and the fallibility of the senses, Plato postulated the existence of entities separate from matter in which a stable truth resides. By adhering to them, the human soul could know the truth.

Plato classifies these entities into two main types:

> -**Mathematical entities** (numbers and figures)
> -**Universal entities** (Ideas or species)

Mathematical entities: These are abstract entities such as numbers and geometric shapes (circles, triangles, etc.). Plato argues that these entities do not exist in the physical world as we perceive them through the senses. For example, when we see a drawn triangle, we are not seeing the "ideal" or perfect triangle that only exists in the World of Ideas. The triangle we observe is an imperfect copy of the ideal triangle.

Universal entities (Ideas or species): These are the Platonic "Forms" or "Ideas." They refer to universal concepts that do not have material existence. For example, the Idea of "justice" or "beauty" cannot be found directly in the material world, but all just or beautiful things participate in that universal idea.

Furthermore, our intellect employs two types of abstraction to reach the knowledge of the truth. One is that by which it apprehends mathematical numbers, magnitudes, and mathematical figures without apprehending sensible matter. It is evident that when we understand the number two or three or a line and a surface, or a triangle and a square, nothing associated with heat or cold or other things similar to those perceived by the senses enters into our apprehension. The other abstraction is the one employed by our intellect when it understands something universally without paying attention to the particular. For example, when we think of man without thinking at all of Socrates, Plato, or any other individual; and the same applies to other things.

This text explains two forms of abstraction performed by the intellect: one in the realm of mathematics, where it abstracts from sensible matter, and another in the realm of universals, where it abstracts from particular individuals.

While mathematical entities may appear in multiple instances of the same species, universal ideas are unique and represent the essence of things. For example, we can draw many different circles, but all of them are imperfect manifestations of the "Idea" or "Form" of the perfect circle, which only exists in the Ideal World. On the other hand, each Idea, such as that of justice, is unique and cannot manifest in as many different forms as mathematical entities. Just things in the world participate in the singular, immutable "justice" that exists in the World of Ideas.

Plato organizes knowledge and reality into hierarchical levels. Mathematics represents a form of knowledge closer to the sensible world, while universal ideas, such as the Good and the One, are the absolute and divine principles from which everything else derives its being and meaning.

Plato places the One and the Good at the top of his hierarchy of Ideas. The One is the supreme principle of unity, and the Good is the divine principle that gives meaning and value to all other Ideas. These are the foundations of all reality, both intellectual and moral. All other beings (secondary gods, separate intellects) participate to a lesser degree in the One and the Good, meaning they derive their existence and knowledge from these supreme principles.

Secondary gods: Plato conceives them as divine entities that, although not the supreme principle, have a connection to the One and the Good. These gods participate in these principles in a limited way and derive their being from them. They are beings that exercise divine functions within the cosmos but are subordinate to the ultimate principle. They are below the One and the Good and are responsible for specific aspects of cosmic order. They have defined roles in the structure of the universe and act as mediators between the One and the sensible world, helping to maintain harmony and balance in material reality.

Separate intellects: These are abstract and immaterial entities that exist as Forms or Ideas in the World of Ideas. They have no physical existence and do not directly participate in the movement of the cosmos. They are intellects that exist independently of matter and possess pure knowledge. These intellects participate in the One and the Good but are not identical to them. They are entities that, being separate from the material, attain a higher form of knowledge, though they still derive their existence from the supreme principle.

Plato organizes the cosmos in a hierarchical structure, linking the movement of bodies and superior souls to intellectual and divine principles.

Superior souls: In Plato, "superior souls" refer to spiritual entities that are closer to the world of Ideas or Forms and possess a more divine nature. These souls are above human souls and have a deeper and more direct understanding of eternal and universal realities. As such, they are said to have "intellectual virtue," meaning a purer and higher knowledge, closer to the One and the Good. They act as agents that influence the order and movement of the cosmos, regulating and maintaining balance in the universe. These souls not only have superior intellectual understanding but are also virtuous due to their participation in the supreme principles.

The movement of bodies is linked to souls. In Platonic thought, bodies do not move on their own but require a principle of movement, which is the soul. In this way, celestial bodies and everything in the cosmos receive their movement thanks to the influence of the soul, which is the source of life and activity.

Plato organizes the cosmos hierarchically. At the pinnacle of this structure is the **supreme soul** (probably the World Soul), which is responsible for moving the celestial bodies. This supreme soul influences all other souls and bodies, establishing a cosmic order where each entity occupies a place according to its level of participation in the supreme principles.

In Plato's philosophy, the concept of the supreme soul is not exactly identified with the notion of God in the traditional theological sense, but it does have a fundamental role that resembles the idea of a supreme divine entity.

The World Soul (or *Anima Mundi*) is an entity that gives life and movement to the cosmos. It is the principle that organizes and harmonizes the universe, infusing it with order and coherence. It acts as the animating and regulating principle of the cosmos. It is responsible for life and movement in the universe, acting similarly to how one might think of a deity that organizes and shapes reality.

The human soul occupies an intermediate position between superior souls and the material world. It is not at the level of divine souls nor the abstract Forms. It has a connection with the World of Ideas through its capacity for knowledge and reason. It can reason and know the Forms, though imperfectly and limitedly compared to superior souls. Intellectual virtue is an essential part of human soul life, as through knowledge and reason, the soul can approach an understanding of the Ideas. Its role in the cosmos is less central than that of the superior souls or the World Soul, but its moral and intellectual life contributes to the universe's order and harmony. Plato maintains that the human soul has the capacity to remember (or *anamnesis*) the Ideas it knew before its incarnation in the material world.

The human soul is embodied in a material body, which limits its perception and knowledge compared to its state in the World of Ideas. Life in the body is seen as a kind of fall or limitation for the soul, which must strive to remember and reach true knowledge. Education and philosophy are viewed as means for the human soul to rise toward true knowledge and free itself from the body's limitations. Philosophy is especially important because it guides the soul in its search for truth and the understanding of Ideas.

Although the human soul does not have the same active role in the organization of the cosmos as the World Soul or the superior souls, its moral and intellectual life impacts the order of its own existence and its relationship with the cosmos. The human soul participates in the cosmic order through its relationship with the World Soul and its effort to align with the Forms. The harmony of the human soul contributes to the general order of the universe.

Plato conceives **the One** as the absolute and original principle of all reality. It is the source from which all things emanate and the foundation of their unity. In this sense, the One is the principle of identity and order in the cosmos.

On the other hand, **the Good** is understood as the ultimate end to which all things aspire. It is the ideal toward which all beings orient themselves and the cause of all that is good and beautiful. The Good is, therefore, the teleological principle of the universe.

Plato establishes a close connection between the One and the Good. He suggests that the One is the condition for the possibility of the Good, meaning that without unity, goodness could not exist. Likewise, he asserts that the Good is the highest manifestation of the One, the way in which the One reveals itself in the world.

Although closely related, the One and the Good are not identical.

The One is understood as a potential, a source from which all things emanate, while the Good is the act of this potential, its concrete manifestation in the world.

The One is often described as a transcendent principle, beyond all determination and category. The Good, on the other hand, is more closely related to the sensible world and human experience, though it always retains its ideal character.

While the Supreme Soul has a role similar to that of a deity in terms of cosmic organization and life, the One or the Good is the most fundamental and abstract principle in the Platonic hierarchy, and it does not completely resemble the idea of a personal God.

In summary, in Plato's conception, the cosmos is an ordered system where superior souls, with their intellectual capacity and virtues, not only govern knowledge but are also the source of movement for bodies, including the celestial ones, under the influence of the Supreme Soul.

Finally, Saint Thomas briefly describes the Platonic conception of bodies.

In Platonic philosophy, **three main types of bodies** can be distinguished based on their nature and relationship with the soul:

1-Celestial bodies, also called aerial or ethereal bodies: These bodies are considered immortal and have a lighter and more subtle nature than earthly material bodies. They are associated with the celestial and spiritual realm and are less dense than earthly bodies. Examples include the stars and other bodies in the sky that move according to a more divine and pure nature. These bodies are not subject to the corruption and decay of material bodies.

2-Bodies free from earthly corporeality: These bodies do not have a material nature in the ordinary sense. They are closer to the divine world and have a more spiritual existence. Example include the "demons" in Platonic philosophy, which should not be confused with the Judeo-Christian conception of malevolent demons; they are spiritual beings that operate in the divine realm and have bodies that are not subject to earthly matter.

3-Earthly bodies: These are bodies with a material nature and are subject to corruption and change. They are physical bodies that undergo birth, growth, illness, and death. Examples include human bodies and other living beings in the material world. These bodies are animated by the soul but are not completely identified with the essence of the soul. In Platonic thought, the soul resides in a perpetual and invisible inner body, distinct from the material body.

Finally, the Platonists attributed moral characteristics to both demons and humans, distinguishing between good and bad people, and good and bad demons. Celestial souls, separated intellects, and gods were always considered good.

Thus, they placed four orders between us and the supreme god: the order of the second gods, the order of separated intellects, the order of celestial souls, and the order of good or bad demons. ***All these intermediate orders, if they truly existed, would be what we know as angels****. Demons are also called angels in Sacred Scripture; and as Augustine asserts in the Enchiridion, the souls of celestial bodies—assuming they are animated—should also be counted among the angels.*

This text presents the idea that, according to Platonic cosmology and theology, the intermediate orders between humans and the One-Good could be considered angels in a broad sense. Demons, being called angels in Sacred Scripture, and celestial souls, would also be included in this category according to Augustine's definition.

The question is, therefore: Did Plato approach the Judeo-Christian metaphysical concept of angels?

Plato's theory does not explicitly mention the idea of "angels" in the way we understand it in Judeo-Christian tradition, but it does present concepts with similarities and parallels to this notion.

THE ANGELS

In Platonic philosophy, superior or divine souls are spiritual entities that are closer to the Ideas or the One. These souls play a role in the organization and order of the cosmos and participate in the hierarchical structure of the universe. They are associated with intellectual virtue and have a direct relationship with the World Soul.

Angels, on the other hand, are spiritual beings that act as messengers or intermediaries between God and humanity. They also play roles in organizing the cosmos and administering divine order. Their function includes guiding and protecting human beings and executing God's will.

There is a similarity here: both concepts involve spiritual beings with roles in organizing the cosmos and in the relationship between the divine and the human.

In Plato, the World Soul *(Anima Mundi)* is the principle that gives life and movement to the cosmos. It acts as an animating entity that organizes and harmonizes the universe.

Although the World Soul does not have a direct parallel in Judeo-Christian tradition, the notion of spiritual beings influencing cosmic order may resonate with the idea of angels as divine agents intervening in the world.

Both concepts reflect the idea of spiritual influence in the organization and order of the universe.

In Plato, superior souls and separated intellects can be seen as mediators between the One and the material world. They have a role in transmitting knowledge and influencing the sensible world.

Angels serve as mediators between God and humans, transmitting messages and executing divine will.

Thus, the function of mediation and influence in world order are common themes in both systems of thought.

However, important differences exist:

1-Nature of spiritual beings: In Plato, the concepts of superior souls and the World Soul are more related to the metaphysical structure of the cosmos and participation in the Ideas. They are not attributed personal or specific roles in human life. In contrast, angels have personal and specific roles in human life, such as guardians and messengers. Their function is clearly defined in terms of direct interaction and assistance with humanity.

2-Focus on divinity: In Plato, the structure of reality and the hierarchy of souls are centered on participation in the Ideas and the One. There is no personal divine figure in the Christian sense. In contrast, angels operate within the context of a personal relationship with God and have a direct role in executing divine will in the world.

2. ARISTOTLE'S OPINION

In Chapter II, Saint Thomas studies and presents the Aristotelian view on the nature of universals, the movement of celestial bodies, and the role of separate substances. Finally, he compares these concepts with the Platonic ideas and those of other thinkers, such as Avicenna.

The primary purpose of the Angelic Doctor is to present Aristotle's perspective on the relationship between the material world and immaterial substances (or intelligences) and how Aristotle sought to prove the existence of an first unmoved mover through his analysis of motion and celestial bodies. Saint Thomas not only presents this position but also offers a moderate critique of it, comparing Aristotle's ideas with other philosophical currents, such as the Platonic, and exploring some limitations of Aristotelian conclusions.

Before continuing, it is important to make a fundamental distinction in Aristotelian philosophy between separate substances and immaterial substances. Although both concepts refer to entities that are not composed of physical matter, their relationship to the material and their mode of existence differ in key aspects.

1-Separate Substances: These are substances that exist entirely independent of matter and any relationship with material bodies. They are called "separate" because they do not depend on the physical world for their existence and are not bound to a body. Aristotle uses this term to refer to entities such as unmoved movers or gods, which are purely intellectual and eternal intelligences. These entities are responsible for the activity of the universe, acting as final causes of motion without being themselves in motion or connected to matter.

2-Immaterial Substances: The concept of immaterial substances is broader and includes all entities that are not made of matter. All separate substances are immaterial since they do not possess a material nature. However, not all immaterial substances are completely separate from matter. Some immaterial substances are linked to bodies, though they are not entirely identified with them.

An example of an immaterial but not completely separate substance is the soul of the celestial spheres. According to Aristotle, the celestial spheres are animated by a soul that is immaterial but acts upon the body of the sphere. Although this soul is immaterial, it is not entirely separate in the Aristotelian sense, as it is linked to a body—the sphere it moves.

Another example of an immaterial substance is the human soul. Although the soul, especially the rational part, is immaterial and does not depend on matter to perform its intellectual functions, in Aristotle's view, it is united with the body during life and acts through it. The separation between soul and body only occurs in death.

The distinction between these two kinds of substances is based on their relationship to matter. Separate substances exist entirely independently of the material world and have no connection to physical bodies. Immaterial substances, on the other hand, can be linked to bodies, as is the case with the human soul or the souls of celestial spheres, although they themselves are not composed of matter.

For a clearer explanation, we will divide the Chapter into three parts:

First part

Saint Thomas describes how Aristotle argues for the existence of an first unmoved mover that is incorporeal, intellectual, and desired as the supreme good by the first mobile, which is a being animated with an intellectual soul.

1-Universals and mathematical things do not exist separately: Aristotle rejects the Platonic idea that universals exist separately from particular objects. He points out that what the human intellect can know separately does not necessarily exist as separate in reality. In other words, universals or essences (such as the concept of "humanity" or "redness") do not exist outside the individuals that manifest them but are the essences of particular things. Likewise, mathematical quantities do not exist apart from sensible objects but are characteristics of those objects.

2-The most reliable way to investigate immaterial substances: Aristotle proposes a more reliable method for studying substances that are immaterial (such as the soul or God): studying motion. This analysis is based on the observation of the sensible world and relies on both reason and experience.

3-The principle that *everything that moves is moved by another*: Aristotle observes that everything that moves does so due to an external cause that moves it. If something appears to move itself, it is actually because one part moves another, as in living beings, where the soul moves the body. Moreover, there cannot be an infinite chain of movers and moved things, as if the first mover fails, the entire system fails as well. Therefore, there must be an first unmoved mover that is not moved by another.

4-The nature of the first unmoved mover: Aristotle concludes that the first unmoved mover must be incorporeal, that is, without body and magnitude. The reason behind this is that no finite force can move eternally. In other words, any material object or limited force cannot generate continuous and infinite motion due to its natural constraints and physical limitations.

Since motion is eternal and has no beginning or end in time, there must be a cause of this motion that is also eternal and infinite. Therefore, the first unmoved mover must be of infinite nature. If it were finite, it would be subject to the limitations and eventual exhaustion of a material body, which would contradict the idea of eternal motion.

Aristotle argues that the first unmoved mover cannot have any properties inherent to an extensive body. This means it cannot have physical properties like magnitude or mass, as a finite body, by its very nature, could not sustain infinite motion.

5-The good as unmoved mover: In the world of mobile beings, Aristotle introduces the idea that the good acts as an unmoved mover. This means that the good is the ultimate cause of movement, even though it does not directly participate in it. According to Aristotle, the good is what moves other beings to act, but it does not move itself. In other words, the good exerts an influence that causes desirous beings to move towards it.

The desirous beings, those who desire or seek something, move toward this good because they perceive it as desirable in itself. The good, therefore, is the ultimate goal or final end toward which all beings tend. Although the good neither moves nor changes, it is the ideal reality toward which all beings are directed and seek to attain. This concept implies that the good is an essential force that guides and motivates movement and desire in the world of mobile beings.

In Aristotle's philosophy, the concept of "good" is not exactly the same as the "unmoved mover," though they are related.

Aristotle introduces the concept of the unmoved mover in his work *Metaphysics*. This unmoved mover is a completely immutable and incorporeal entity that causes motion without being moved by anything else. Aristotle sees it as necessary to explain the existence of eternal movement and the chain of causes. The unmoved mover is, in essence, a primary cause that triggers motion in the universe without being subject to it.

The "good" in Aristotelian philosophy, especially in his ethics and metaphysics, refers to what is desirable in itself and what gives purpose to actions and desires. In the context of the unmoved mover, the good can be considered a final cause that directs beings towards their perfection and purpose. Although Aristotle does not directly identify the good with the unmoved mover, the good functions similarly in that it acts as a goal toward which beings move.

Consequently, the good is not the unmoved mover in the strict sense according to Aristotle. The unmoved mover is an abstract entity fundamental to explaining eternal movement. The good, on the other hand, is a final principle that motivates and directs actions toward perfection. Although the two concepts are interrelated, the good is not the unmoved mover as the primordial cause of movement, but rather a final cause that provides direction and purpose.

6-The primacy of the intellect in desire: Aristotle distinguishes between different types of desire in beings: intellectual desire and sensory desire.

Intellectual desire is the desire based on the intellect, which is capable of understanding the good in its purest and most absolute form. This desire seeks not only what appears good but what is intrinsically good in absolute terms. In other words, the intellect can grasp the absolute good or perfection, allowing desire to be directed toward what is truly good.

On the other hand, sensory desire is tied to the senses and emotions. This desire is limited to wanting what appears good based on immediate perceptions and physical needs, without reaching an understanding of the absolute good or perfection itself.

Since the absolute good can only be grasped by the intellect and not by the senses, the desire to reach and comprehend the unmoved mover, which represents the good in its purest and most perfect form, must be guided by intellectual desire. This is because the unmoved mover is the final cause of movement in the universe, and its absolute perfection demands a desire that can only be expressed through the intellect.

As a result, intellectual desire is superior to sensory desire because it can desire the good in its purest and most absolute form. Since the unmoved mover represents that absolute good, the desire toward it must be intellectual, reflecting the idea that appreciating and desiring the absolutely perfect requires an intellectual capacity that transcends sensory perceptions.

7-The first mover (or first moved mover) as a being with will and intelligence: The first unmoved mover is the immutable and pure principle of movement. It is not subject to movement or change and is the ultimate cause of movement in the universe. The first mover is the first heaven or the first celestial sphere, which moves eternally due to the influence of the first unmoved mover.

The first mover, although not the unmoved mover itself, desires (appetizes) the unmoved mover with intellectual desire. This means that the first mover is oriented toward the unmoved mover as the supreme good.

The "intellectual appetite" here should not be interpreted literally in a human sense, but rather refers to an orientation or direction toward the unmoved mover as the final and perfect cause of movement. The first mover follows the cause of the unmoved mover because it perceives (in metaphorical terms) the unmoved mover as the absolute good.

The unmoved mover is the cause of the first mover's motion, and hence, the first mover reflects a kind of "desire" or aspiration toward the unmoved mover, which is the source of eternal motion.

This first mover is not a passive being but one that possesses the qualities of will and intelligence.

For any being to desire or intellectually crave the absolute good (which is the unmoved mover), it must have both will and intelligence. Will is the faculty that allows a being to desire or want something, while intelligence enables it to understand and recognize the good in its purest form. In other words, a being capable of desiring what is intrinsically good and perfect must possess these two fundamental capacities.

And since nothing moves except a body, it must be concluded that the first mover is a body animated by an intellectual soul.

Second Part

In this second part, St. Thomas continues to explain Aristotle's ideas about the movement of celestial bodies and the nature of intellectual substances that influence this movement.

1-Eternal movement of celestial bodies: Aristotle holds that not only the "first mover" (the first heaven) moves eternally, but so do the lower spheres, meaning all celestial bodies. For Aristotle, each of these celestial bodies has its own soul, which means they are animated. Each also has its own "desirable object," a goal towards which their movements are directed. This suggests that there are many immaterial substances, that is, intellectual entities not united to bodies (separate substances), in addition to the souls that are linked to the celestial bodies.

2-Number of immaterial substances: Aristotle tries to determine the number of these immaterial substances based on the number of celestial movements. Avicenna, one of his followers, offers a different interpretation: instead of counting the movements, Avicenna counts the planets and other higher spheres, such as the sphere of stars and the sphere without stars.

3-Hierarchy of the cosmos: In the Aristotelian system, the cosmos is organized into a hierarchical structure where each element has a specific role and is subordinated to a higher level. This hierarchy is based on the existence of a supreme heaven or first heaven, which acts as the first moved mover responsible for the movement of all other heavens.

Aristotle identifies a first heaven, which is the first mover. This first heaven initiates and maintains the movement of all celestial bodies and is the highest being in the cosmic hierarchy. Its role is essential for the functioning of the cosmos, as its eternal movement provokes the movement of the lower heavens.

All celestial bodies, such as spheres and planets, are organized under the supreme heaven or first mover (or first moved mover). The first heaven, acting as the first mover, regulates and coordinates the movement of the lower spheres and celestial bodies in the system. Each celestial body and sphere depends on this first heaven for its movement and position.

Similarly, the separate substances, which are immaterial entities such as gods or pure intelligences, are hierarchically ordered under the first separate substance. Aristotle identifies this first separate substance as a god (lowercase), which is the final and primary cause in the system. The other separate substances depend on this supreme entity for their existence and function within the hierarchy.

The souls that animate the celestial spheres are also subordinated to the soul of the first heaven. These souls are responsible for the movement of the celestial bodies, but their existence and function are hierarchically subordinated to the soul of the first heaven. The structure of the cosmos is organized in such a way that all celestial souls and bodies are interconnected and depend on the supreme heaven for their movement and existence.

The Aristotelian system establishes a hierarchy in which the supreme heaven, as the first moved mover, is the fundamental entity that causes and regulates the movement of all celestial bodies and separate substances. This hierarchical order ensures that each level has a specific function and depends on the higher level for its existence and movement.

4-The sublunar world: According to Aristotle, in the sublunar world (that is, the world beneath the celestial spheres, including Earth), the only animated bodies are those of animals and plants. He does not admit that the simple elements (such as air or fire) are animated, as they do not possess the necessary organs, like the sense of touch, which is required for any animated being. Therefore, Aristotle does not place animated bodies between us and the celestial bodies.

5-Twofold order of intellectual substances: Aristotle proposes that between humans and the supreme god, there are only two types of intellectual substances: the separate substances and the souls of the spheres.

6-Comparison with Plato's view: Although Aristotle's position seems plausible because it aligns more with what the senses perceive, St. Thomas notes that it has some limitations. For example, there are phenomena that Aristotle cannot adequately explain, such as extraordinary manifestations in people possessed by demons or the actions of magicians. These manifestations seem to require the intervention of intellectual substances that Aristotle does not contemplate in his system.

7-Attempts by scholars to explain Aristotle: Some of them, such as Porphyry, tried to attribute these phenomena to the influence of celestial bodies. Porphyry suggests that certain constellations can influence magicians, enabling them to perform wondrous feats. Additionally, under the influence of the stars, the possessed can foretell future events, as celestial bodies might influence natural dispositions.

St. Thomas criticizes the idea that these phenomena can be explained by the influence of celestial bodies. He argues that there are manifestations that clearly cannot be attributed to physical or astronomical causes, such as the possessed speaking of knowledge they have never learned, using languages they do not know, or magicians creating images that give answers or move. These actions cannot be explained solely by the influence of celestial bodies.

8-The Platonist view: According to the Platonists, these unusual manifestations can be better explained by assuming the intervention of demons. The demons would be responsible for these extraordinary phenomena, offering a more adequate explanation than astronomical influence.

Third Part

In this third part, St. Thomas delves deeper into the argument concerning the relationship between immaterial substances and the number of celestial movements, questioning Aristotle's position and offering a critical perspective.

He argues that limiting, as Aristotle does, the number of immaterial substances based on the number of celestial movements is unreasonable, since higher beings should not be subordinated to lower ones, but rather the opposite. Moreover, the greatness and power of higher substances cannot be established merely by observing the lower ones, as immaterial substances surpass celestial bodies in a much more significant way than celestial bodies surpass the elemental bodies (bodies constituted by the basic elements of nature, according to the ancient conception of matter: air, water, earth, and fire).

Aristotle holds that in the heavens, all movement has an end, and that the superior substances, being the most optimal, are the ends of those movements. However, it is objected that the number of immaterial substances does not necessarily have to correspond to the number of celestial movements. Here, a distinction is introduced between proximate and remote ends: the proximate end of the movement does not have to be the supreme immaterial substance (God), as there may exist other orders of intermediate substances between the celestial body and the first immaterial substance.

Thus, Avicenna proposed that the immediate end of celestial movements is not the first cause (God), but an intermediate first intelligence. This implies that the number of immaterial substances could be greater than the number of celestial movements, and although Aristotle proposed his thesis as reasonable, he did not consider it definitive, acknowledging his own limitations on the matter.

Finally, a possible objection is addressed, which might arise from the fact that Aristotle bases his argument on the eternity of movement, which contradicts the Christian faith. However, it is clarified that the value of his argument does not diminish even if one disregards the eternity of movement, since the uniformity of celestial movements also indicates that the first mover (God) possesses the ability to always move uniformly and constantly, leading to the same conclusion.

Aristotle and angels

Aristotle does not directly contemplate the existence of angels as understood in the Judeo-Christian tradition. His cosmology and metaphysics focus on concepts such as the unmoved first mover and separate substances, but he does not address the idea of angels. Below, I explain how his thought relates to the idea of spiritual beings and whether there are approximations to the notion of angels.

1-The first unmoved mover: Aristotle introduces the concept of the first unmoved mover as the ultimate principle of movement in the universe. This first mover, which causes movement without being moved, is not a personal or spiritual being but rather a metaphysical entity that ensures the order and movement of the cosmos. According to Aristotle, the first unmoved mover is incorporeal and devoid of magnitude, meaning it does not possess a physical body or material characteristics. Thus, it is not an angel.

2-Souls of celestial bodies: Aristotle also speaks of the souls of celestial bodies, which animate and move the stars. However, these souls do not have the same spiritual connotation as angels. The celestial souls in Aristotelian thought are principles of movement and organization rather than personal beings with will and spiritual purpose.

3-Hierarchy of substances: Aristotle establishes a hierarchy of substances in the cosmos, where the separate substances (like the first unmoved mover) are at the top, and the celestial bodies, animated by their own souls, are subordinated to these superior substances. This hierarchical order includes a set of immaterial beings, but they are not necessarily equivalent to angels.

4. Approximation to the idea of angels: Although Aristotle does not directly address the idea of angels, his concepts of immaterial beings and separate substances bear some resemblance to the notion of spiritual entities playing roles in the universe. However, it is important to note that Aristotle does not develop a theory of personal spiritual beings with intelligence and will, such as angels. Instead, he focuses on a cosmology and metaphysics that explain the order and movement of the cosmos through abstract principles and non-material entities.

Conclusion

Although Aristotle does not contemplate the existence of angels as understood in the Judeo-Christian tradition, his ideas about the first unmoved mover and separate substances present a metaphysical structure that may appear similar in terms of immaterial beings. However, the differences are significant, as Aristotle does not introduce a conception of personal spiritual beings with specific roles in the order of the universe.

3. HARMONY AND DISCORD: ARISTOTLE AND PLATO ON IMMATERIAL SUBSTANCES (1)

In Chapter III, St. Thomas examines the similarities and differences between Plato's and Aristotle's doctrines regarding immaterial substances. He focuses his study on the following points:

1-The mode of existence of immaterial substances: Plato asserts that all immaterial substances derive their unity and goodness from the first being: the One. This first being is the cause of the unity and goodness of the lower substances, much like how the sun is the cause of the brightness of the air that participates in its light.

Aristotle also maintains that what is most truly and fundamentally a being (the unmoved mover) must be the cause of being and truth for all other things.

At this point, both philosophers agree that the first being is the ultimate cause of all immaterial substances.

2-Nature of immaterial substances: Both Plato and Aristotle believe that immaterial substances are completely free from matter, but not from all potentiality. This means that, although they lack physical bodies and material substance, they are still composed of potentiality and act, since participation in something implies a composition of these principles. The first being, on the other hand, is pure act, without any mixture of potentiality, as it does not participate in anything but is self-sufficient.

Aristotle reaffirms this by noting that the first and most true being is pure act, while other things, which do not reach this degree, have some mixture of potentiality.

3-Providence: Plato teaches that the One, being the Good itself, takes care of all lower things through providence. This supreme being or first god is the principle and source of all that is good. Its goodness influences and extends to all that exists. Providence is how this first god, as the Good, takes care of and governs all lower things. "Lower things" refers to all entities that are below this god in the hierarchy of beings, including both spiritual beings (such as separate intelligences) and material beings (such as humans and nature). Higher entities (those beings closest to the first god, such as separate intelligences in the Platonic hierarchy) **participate** in this goodness, meaning they receive a share of it or are influenced by it. By doing so, these higher entities also take on the responsibility of providing for and governing the lower entities. Thus, a **hierarchy** is established where each higher being is responsible for aiding, caring for, and ruling over the lower beings. The first god is the source of all goodness, but its providence descends through this hierarchy, with each higher level governing and guiding the lower levels, creating an orderly system where each being has a role and function within the structure of the universe.

Aristotle, although not strictly following the same notion, agrees that there is a separate and good being that provides for all, and that higher orders are more perfect and less defective than the lower ones. For example, human souls are more prone to deviation from truth and goodness compared to higher intelligences.

Aristotle conceives a supreme being or unmoved mover, which is separate from the physical world, meaning it is not mixed with matter or subject to change. This being is good in the sense that it is pure act, without potentiality, making it perfect and the ultimate cause of movement, being, and order in the universe.

Although Aristotle does not precisely follow Plato's idea of the One or Good, he agrees that there exists a transcendent being that acts as a source of perfection and is the cause of order in the cosmos. This being, being completely actual and without defects, is removed from any type of imperfection.

This supreme being, while not intervening directly in the world as a personal god (as seen in Plato's providence or Christian theology), is the ultimate cause of everything that happens. All things in the universe are drawn towards it because it is the most perfect being, the ultimate end towards which all beings tend, although this care is indirect.

This being does not actively concern itself with the details of what happens in the world (unlike Plato's idea of providence or Christian theology), but its mere existence establishes a hierarchical order in which the most perfect beings resemble this supreme being more closely and are less prone to imperfection.

Aristotle argues that the universe is ordered in a hierarchy of beings, with higher orders (such as separate intelligences) being more perfect and thus closer to the supreme being. These higher beings have fewer defects than the lower beings due to their proximity to the unmoved mover.

For example, human souls, being in a lower order, are further from the supreme being and thus more prone to deviation from the good and the true. This means that humans, due to their connection with matter and imperfections, can more easily err and have difficulties in reaching perfect knowledge and goodness.

Unlike higher intelligences (which are purely intellectual and not subject to the limitations of matter), human souls have a complicated relationship with truth and goodness. Being at a lower level in the hierarchy of beings and being linked to a material body, human souls can stray from true knowledge and goodness.

This means that humans, due to their mixed nature (soul and body), are more subject to making errors in their understanding of reality and deviating morally. Higher intelligences, on the other hand, do not have these limitations and are thus always in harmony with truth and goodness.

4-The good and the truth: Both Plato and Aristotle link the Good and the Truth to pure act, but from different approaches. For Plato, pure act is the perfection of the Ideas or Forms. Goodness and truth are found in these perfect Forms, and the objects of the sensible world attain goodness and truth by participating in them. The Supreme Good, as the Form of the Good, is the pure act that guides everything towards perfection.

For Aristotle, pure act is the unmoved mover, the perfect reality that causes movement and order in the universe. Here, the Good and the Truth are related to the natural purpose or end of beings. Something is good if it fulfills its natural end and true if it reflects its essential nature.

What you should remember

1-Participation and causality: Both philosophers use the concept of **participation** to explain the relationship between immaterial substances. However, Plato's notion of participation has a more ontological character, while in Aristotle it is more linked to efficient causality.

For Plato, participation has an ontological character because it refers to how entities in the sensible world (material objects) acquire their properties or qualities from the Ideas or Forms. The Ideas or Forms are abstract and perfect entities that exist on a higher level than the material world. Material objects participate in the Ideas. For example, a particular object, like a triangle drawn in the sand, participates in the Idea of the triangle. Participation here means that the material triangle reflects (or shares) the form and properties of the Idea of the triangle, although imperfectly. This type of participation explains the existence and characteristics of material objects in terms of their relationship with the Ideas. It is a matter of being and essence: material objects are what they are (their "being") because they participate in the Ideas, which are the true and eternal reality.

Aristotle, on the other hand, conceives participation in terms of efficient causality. Efficient causality refers to how one entity causes or produces another entity. In Aristotle's case, participation is linked to how entities in the material world receive their form and existence from causes. For Aristotle, participation means that material things have their existence and characteristics due to the influence of efficient causes. For example, a chair is what it is due to the carpenter's skill (the efficient cause) and the design (the form). Material substances are caused and formed by a combination of matter and form, and this formation is a process of causality. Rather than viewing participation as a matter of being and essence as in Plato, Aristotle sees it in terms of how material entities are caused by or depend on other entities or causes. Here, participation means that something receives its being and qualities as a result of its relation to other causes or principles.

2-Potency and act: The distinction between potency and act is fundamental in both philosophies. However, Aristotle develops this distinction more systematically and applies it to a broad range of phenomena, while in Plato this distinction is more linked to the theory of the Forms.

Aristotle introduces the distinction between potency and act as a central part of his philosophy. This distinction helps explain how beings change and develop.

Potency refers to the capacity of a being to change or receive a form. It is the potentiality something has to acquire a certain form or state. For example, a seed has the potency to become a tree, even though it is not yet a tree.

Act is the actualization of that potentiality. It represents the state in which something has achieved its final form or purpose. Following the previous example, when the seed grows and becomes a tree, it has actualized its potency into act.

Aristotle applies this distinction to a wide range of phenomena, not only in biology or physics but also in metaphysics. The distinction between potency and act allows him to explain how objects move from a potential state to an actual state, how they develop, and how their changes and movements are understood.

In Plato, the distinction between potency and act is linked to his theory of the Forms. Plato considers that the sensible world is an imperfect copy of a world of perfect and immutable Forms. The Forms are ideal and eternal realities that serve as models for material objects.

In Platonic philosophy, the distinction between potency and act is not as central or systematic as it is in Aristotle. Instead of focusing on this distinction, Plato concentrates more on the relationship between the sensible world and the world of the Forms. The distinction between potency and act in Plato is reflected in the way material objects attempt to imitate or participate in the perfect Forms, but the theory of the Forms does not explicitly use these concepts in the same way as Aristotle.

3-Providence and hierarchy: The idea of a cosmic providence is common to both philosophies. However, Aristotle presents a more teleological view, where purpose is inherent in natural beings, while in Plato, purpose is more related to participation in the Good.

In Plato, providence is related to the idea of the Supreme Good, or the One, which is the source of all goodness and order in the universe. The Supreme Good oversees lower beings through a kind of hierarchical order and participation.

Entities in the universe obtain their purpose and order by participating in the Supreme Good. Each level of the cosmic hierarchy (separate understandings, souls, material bodies) receives its guidance and purpose from how it participates in this ultimate source of goodness.

In this view, hierarchy is based on the relationship of participation in the Good. Higher beings, being closer to the Supreme Good, have greater perfection and the ability to influence lower levels. Lower beings, in turn, benefit from the order and purpose derived from higher beings.

Aristotle also accepts the idea of a cosmic providence, but his approach is different. Instead of focusing on participation in the Supreme Good, Aristotle develops a more teleological view, where purpose is inherent in the nature of beings.

In Aristotelian philosophy, providence is linked to the notion of teleology, that is, the idea that natural beings have an intrinsic purpose or end. Aristotle believes that each entity in the universe has a purpose or end that guides its development and existence.

Hierarchy in Aristotle is based on the teleological nature of beings. Higher beings have a more complete and perfect purpose, while lower beings have less developed ends. Hierarchy is determined by the complexity and degree of realization of each being's natural purposes.

4-Good and truth: Both Aristotle and Plato link the good and the truth with the pure act. However, Aristotle emphasizes the relationship between the good and purpose, while Plato links it to participation in the Forms.

In Plato, the pure act refers to reality in its most perfect and complete form, which is the essence of the Ideas or Forms themselves. The Ideas, such as the Supreme Good, represent the pure act in its highest expression, as they have no imperfection or potentiality.

For Plato, the good and the truth are found in the Forms. Objects in the sensible world acquire the good and the truth through their participation in these perfect Forms. For example, an action is good insofar as it participates in the Form of the Good, and knowledge is true insofar as it participates in the Form of Truth.

Participation in the Forms implies that the good and the truth are manifestations of the pure act, which is the complete realization of qualities in the Ideas. The Supreme Good, as the Form of the Good, is the pure act that directs everything else towards perfection and truth.

Aristotle also uses the concept of the pure act to describe absolute perfection. For him, the pure act is the unmoved mover, the being in a state of complete reality without potentiality, and which causes movement and order in the universe.

In Aristotelian philosophy, the good and the truth are intimately related to the intrinsic purpose or end of beings. The good is defined in terms of the realization of something's natural purpose. For example, a being is good if it fulfills its natural purpose, and truth is manifested when something fulfills its purpose and reflects its essential nature.

For Aristotle, the pure act is not only the source of perfect reality but also the source of purpose in the universe. The good and the truth are found to the extent that beings fulfill their natural purpose and achieve their full realization.

Conclusions

In Chapter III that we analyzed, Saint Thomas shows that, despite their differences, Aristotle and Plato share a common view on the nature of immaterial substances and their place in the cosmos. Both philosophies postulate a hierarchical and ordered reality, where superior substances exercise a providential function over the inferior ones. However, differences arise in how they conceive of participation, causality, purpose, and the relationship between the good and the truth.

4. HARMONY AND DISCORD: ARISTOTLE AND PLATO ON IMMATERIAL SUBSTANCES (2)

In Chapter IV, Saint Thomas delves into the divergences between Aristotle's and Plato's conceptions of immaterial substances.

The following points reflect significant differences in how Plato and Aristotle understand and organize immaterial substances, celestial hierarchy, and the role of divine entities in their philosophies.

1-Hierarchy of immaterial substances. Plato places a dual order of immaterial substances above the souls of celestial bodies. These are the intelligences (or noetic beings) and the gods (separate intelligible species).

In Platonic philosophy, immaterial substances are those that exist independently of the physical world. This includes entities not subject to the changes and limitations of the material world, such as souls and Ideas. In this case, Plato proposes a hierarchy of substances that are "above" the souls of celestial bodies.

Intelligences or noetic beings (also known as *nous* in Greek thought) are beings capable of knowing and understanding divine Ideas. Plato argues that intelligences participate in the divine Ideas, meaning they acquire their capacity to understand through their relationship with the World of Ideas or Forms. For Plato, Ideas are eternal and perfect realities, and intelligences have access to these through a kind of intellectual intuition or participation.

Intelligences understand through participation in the divine Ideas, and the gods, as intelligible species, are principles that enable intelligences to know.

The gods in this hierarchy should not be understood as anthropomorphic beings but as intelligible species, i.e., principles or models that allow intelligences to know the intelligible world. In Platonic thought, Ideas or

Forms are themselves intelligible (can be understood), and the gods mentioned here are eternal principles that make the knowledge of intelligences possible.

Intelligences depend on the gods (separate intelligible species) to be able to know. The gods, in this context, are principles that transcend the intelligences and make the process of knowledge possible. That is, intelligences could not know if it were not for the existence of these higher principles, which provide them with the ability to comprehend the intelligible world.

In conclusion: The hierarchical structure of the intelligible world in Platonic philosophy reveals that intelligences participate in Ideas and are subordinate to higher principles, which Plato identifies as "gods" or separate intelligible species. These intelligences acquire knowledge through their participation in such principles, which make the act of knowing possible.

Unlike Plato, Aristotle does not accept the existence of separate universals or transcendent Ideas. For Plato, universals (or Ideas) exist in a separate world, and particular things participate in them. In contrast, Aristotle holds that universals do not exist apart from individual things; that is, the concept of a thing is inseparably tied to the thing itself in the physical world. This means there is no "dual order" of separate realities, as Plato imagined (a sensible world and an intelligible world).

Aristotle places the supreme god, also known as the unmoved mover, at the top of the order of substances. The unmoved mover is the ultimate cause of all movement in the universe but does not move or change itself. It acts as the end or goal toward which everything tends, without needing to be moved by another.

The Stagirite describes the unmoved mover as intelligent and also as the object of its own intelligence. This means that God (the unmoved mover) does not know or understand through participation in something external, but its act of knowing is based on its own essence. God is pure act, always contemplating its own perfection and, therefore, does not depend on anything outside itself to know.

Apart from the unmoved mover, there are other immaterial substances, which Aristotle describes as subordinate to this supreme entity. These substances are also intelligent, but unlike the unmoved mover, they are not self-sufficient in their act of knowing. They attain understanding (the capacity to comprehend) through participation in higher substances, meaning their knowledge depends on something higher than themselves, unlike the self-sufficiency of the unmoved mover.

In Aristotle, there is no participation in the Platonic sense. Lower substances participate in a broader sense, directing their intellectual activity toward higher substances, such as the unmoved mover, which is the ultimate end of knowledge and the final cause that moves them. Subordinate intelligent substances achieve their intellectual perfection and understanding by virtue of their relationship with the unmoved mover or with higher substances in the hierarchy of being.

Aristotle presents a more unified view of the world, where knowledge and being are linked to the substances themselves without resorting to a separate world of universals. The unmoved mover is the supreme being, and its knowledge is completely self-sufficient, while other intelligent substances depend on their relationship with the higher to know.

2-Number of separate intelligences. Plato does not limit the number of separate intelligences to the number of celestial movements. His theory is based on the nature of things themselves, beyond movements.

Aristotle limits the number of immaterial substances to the number of celestial movements. His approach is more practical and related to the number of celestial spheres in the cosmos.

Plato's theory of intelligences is not conditioned by the physical cosmos but is grounded in **the nature of things themselves**. Intelligences, or separate intelligences, are spiritual principles that do not depend on the movement of the stars or the physical structure of the universe.

Unlike theories that associate the existence of intelligences with the movements of celestial bodies (such as Aristotle's, where each celestial movement is linked to an intelligence that guides it), Plato does not limit the number of separate intelligences to the number of movements in the cosmos. According to Aristotle, each celestial sphere (the orbits of planets and stars) is moved by a particular intelligence, establishing a direct correspondence between the intelligences and the number of celestial movements.

Plato's Ideas and separate intelligences are fundamental principles that exist beyond any physical entity, and their number is not subject to the limitations of the sensible world or moving bodies.

Plato grounds his theory in the nature of intelligences as independent entities not tied to the material cosmos or astronomical movements. Therefore, the number of separate intelligences does not have to match the number of celestial movements; it is determined by the structure of the intelligible world and not by the physical arrangement of the universe.

For Aristotle, the number of immaterial substances is strictly limited by the number of celestial movements. Each of these intelligences is associated with a celestial sphere, that is, with one of the orbits that celestial bodies follow in the cosmos. Aristotle conceived the universe as composed of a series of concentric spheres that move planets and stars along their paths.

Each of these celestial movements is driven by an intelligence, which is the cause of the perpetual movement of the corresponding sphere. In this sense, there is a direct and necessary relationship between the number of movements in the cosmos and the number of intelligences.

The unmoved mover is the final cause of all movement in the universe. It does not directly move the celestial spheres but attracts them as the object of desire, causing their movement without being moved itself. Besides the unmoved mover, each celestial sphere is driven by a subordinate intelligence that moves the particular sphere to which it is associated.

As a result, the number of subordinate intelligences is directly linked to the number of spheres in the universe, as each one needs an intelligence to account for its eternal movement.

Aristotle adopts a more practical and cosmological approach compared to Plato. His system is based on the observation of the physical world, and therefore relates immaterial substances to the number of celestial spheres, which correspond to the visible planets and stars in the cosmos.

While for Plato, separate intelligences have an existence in the intelligible world, detached from the physical cosmos, Aristotle attempts to link intellect to observable physical reality. Thus, the number of immaterial substances is limited by the elements that form the visible cosmos.

In Aristotle's view, the universe has a hierarchical and ordered structure, where each part of the cosmos has an efficient cause that explains its movement. This structure includes a finite number of celestial spheres and, therefore, a finite number of intelligences. Each sphere is driven by a specific immaterial substance corresponding to a particular celestial body, creating a closed and well-defined cosmic order.

3-Presence of intermediate souls. Aristotle does not mention intermediate souls or demons in his system. His focus is more on the unmoved mover and inmaterial substances without intermediaries in the hierarchy.

Plato organizes the cosmos into various levels of existence, from the divine and intelligible world to the human and material world. In this scheme, **intermediate souls** occupy a crucial place. These souls are neither as exalted as the souls of celestial bodies (which are closer to the divine world) nor as material as human souls, which are more immersed in the physical world.

Plato refers to these intermediate souls as **demons** (in Greek, δαίμονες, *daímones*), and their function is to act as mediators between the gods and humans. These beings are not entirely divine but are also not subject to the limitations and corruptions of the material world like human souls.

In Platonic cosmology, demons play a crucial role in the communication and connection between higher and lower levels of reality. They act as intermediaries between the divine world and the human world, helping to transmit divine influences, ideas, and messages to human beings. In a sense, they facilitate the flow of knowledge and energy between the divine and material worlds.

They have a spiritual nature, and their function is to assist in the relationship between humans and higher entities, such as gods and separate intelligences. By introducing the category of demons, Plato underscores the need for a bridge between the divine and intelligible world and the sensible and human world, which are so different in their nature.

The souls of celestial bodies (such as those of stars and planets) occupy a place closer to the divine world, and their function in Platonic cosmology is higher than that of human souls. These souls, which guide the movements of celestial bodies, participate more directly in the world of Ideas, being more perfect and eternal.

Intermediate souls or demons are situated between these celestial souls and human souls. Although they are not as sublime as the souls of celestial bodies, their mediating function places them in a privileged position. They help humans, who are more immersed in the sensible and material world, to access the divine and receive celestial influences.

According to Plato, human souls are trapped in the body and, therefore, are further removed from the divine and intelligible world. Demons, by acting as intermediaries, play an important role in helping human souls ascend toward the knowledge of the Ideas and divinity. This spiritual mediation facilitates communication between the human world and the divine, helping humans to achieve a higher level of understanding and spirituality.

While Plato posits the existence of demons as mediators between the gods and humans, Aristotle does not recognize any type of intermediate entity in his philosophy. Instead of relying on intermediaries to explain the connection between the divine and the material world, Aristotle bases his system on a more direct and logical order.

The central concept in Aristotelian cosmology is the unmoved mover, which is the ultimate cause of all movement in the universe. This unmoved mover does not require intermediaries to exert its influence over the cosmos. According to Aristotle, the unmoved mover moves everything in the universe as an object of desire or absolute perfection, without needing to interact directly with lower realities. All movement in the cosmos results from the attraction that more perfect substances exert toward less perfect ones, but without the mediation of intermediate spiritual entities.

Aristotle speaks of intelligences linked to the movement of the celestial spheres. Each celestial sphere, corresponding to a body in the cosmos, is moved by an intelligence. However, these intelligences are not intermediaries between the unmoved mover and humans but have a direct function in the structure of the cosmos.

Intelligences are responsible for the movement of the celestial spheres, but they do not fulfill a spiritual mediation role like the demons in Plato's system. These substances are more elevated than the human world but do not act as transmitters of divine influences between the unmoved mover and humans.

Aristotle adopts a more direct and structured approach, without the need for spiritual intermediaries between different levels of reality. In his cosmology, all movement and order in the universe can be explained through the unmoved mover and the intelligences that move the celestial spheres. There is no need to postulate the existence of intermediate spiritual entities like demons to connect the divine with the human.

Aristotle's unmoved mover is entirely self-sufficient and does not need intermediaries to cause movement in the cosmos. This unmoved mover is pure act, without potentiality, and does not require any form of participation or mediation to exert its influence. Being perfect in itself, it moves the universe simply by being an object of desire and perfection, without interacting directly with the material world.

Conclusion

While both philosophers acknowledge the existence of immaterial realities, their conceptions of these differ significantly in terms of their nature, function, and relationship with the sensible world.

Plato: His view is more dualistic, radically separating the World of Ideas (perfect and unchanging) from the sensible world (imperfect and changing). Separate substances are the archetypes of all that exists, and human understandings participate in them to know reality.

Aristotle: His view is more realistic, integrating the immaterial and the material into a single reality.

5. MATTER AND FORM IN SPIRITUAL SUBSTANCES: THE CASE OF AVICEBRON

In Chapter V, Saint Thomas analyzes Avicebron's view on the nature of substances and their composition.

Avicebron, in his work *Fons Vitae*, maintains that all substances, including spiritual ones, are composed of matter and form. While Plato and Aristotle reserved the combination of matter and form exclusively for material beings, Avicebron extends this principle to spiritual realities as well.

According to this critique, Avicebron's error lies in applying the idea of matter and form composition, which is characteristic of physical objects, to spiritual substances. For Avicebron, both material and spiritual beings share the same ontological structure: both would be constituted by a matter (as a principle of potentiality) and a form (as a principle of determination or actualization).

In contrast, Plato and Aristotle argue that spiritual substances, such as souls or intelligences, do not require matter to exist. Aristotle asserts that only material beings are composed of matter and form, while spiritual substances are pure forms, without matter. Plato, for his part, considers that spiritual realities are independent of matter and belong to the World of Ideas or abstract Forms.

Avicebron also confuses the concept of potency (the capacity to receive a form) with the concept of subject (the one who has a form). This leads him to a misinterpretation of how spiritual substances are composed.

In philosophy, especially in Aristotelian thought, potency refers to the capacity of something to receive or assume a form. It is the principle of possibility, what can come to be or change by receiving a new form. For example, wood (potency) can become a table (form).

The subject, in this context, is the bearer of an already realized or actualized form. It is the entity that possesses a specific form. In the case of spiritual beings, it would be the entity that has a form without the need for matter.

By confusing these two concepts, Avicebron incorrectly interprets the composition of spiritual substances. Instead of recognizing that spiritual substances are pure forms (according to Aristotle and Plato) and do not need matter to exist, Avicebron suggests that these substances have a material potency, like physical beings. In his view, spiritual substances would also be composed of matter (potency) and form, leading to the erroneous idea that immaterial beings (such as souls or intelligences) require a material structure to exist.

The confusion lies in thinking that spiritual substances need a material potency to receive a form, when, according to the Platonic and Aristotelian tradition, spiritual substances are already complete forms and do not require matter to exist. This misunderstanding causes Avicebron to treat spiritual substances similarly to physical objects, which contradicts classical conceptions.

Avicebron analyzes different types of substances and how they are composed. He examines artificial substances (such as man-made objects) and natural substances (the four elements and celestial bodies), and tries to classify the matter and forms of these different types of substances.

In this sense, he distinguishes between:

-**Universal natural matter**: This is the matter of the classical four elements (earth, water, air, and fire) that constitute material objects in the physical world. This matter is characterized by its contrary qualities, such as cold and heat, dry and wet, which differentiate it and cause objects to change and transform in nature.

-Matter of celestial bodies: Avicebron proposes that celestial bodies, such as stars and planets, also have matter, but it is of a superior nature compared to terrestrial elements. This matter is not subject to the same contrary qualities (such as cold or heat) as terrestrial matter, making it more stable and perfect. In this sense, celestial bodies would be immutable or less subject to change than terrestrial bodies.

-Spiritual matter: In a more radical step, Avicebron extends the concept of matter to the spiritual world, asserting that even spiritual substances (such as angels or souls) are composed of matter and form. This "spiritual matter" would be of a different nature from material bodies, but would still exist as a principle of potentiality or receptivity in immaterial substances, making them similar, in a way, to physical beings.

In his view, the diversity in the material world is due to the difference between matter and form that constitute physical beings. Therefore, he applies this principle to the spiritual realm, suggesting that there must be an analogue to corporeality (which in material beings is sustained by matter) also in spiritual substances. In other words, "spiritual matter" would be the basis upon which spiritual forms are actualized, similar to how physical matter supports form in material objects.

Avicebron also compares the degrees of substance in the physical world with those in the spiritual world. In the material world, beings vary in degrees of perfection, from the simplest objects to the most complex. Similarly, Avicebron argues that there are also degrees of superiority and inferiority among spiritual substances, implying that they must be composed of something (spiritual matter) that allows for this variation.

Avicebron asserts that no created substance can be completely simple, as all creation must differ from its Creator. God, in his conception, is the only completely simple being, not composed. Since creatures cannot share this simplicity, they must be composed of matter and form. This is crucial for Avicebron: every created substance has some form of composition, which distinguishes them from the Creator. Therefore, both created physical and spiritual substances must have a dual composition.

Finally, Avicebron considers that spiritual substances are finite due to their form. In scholastic thought, the form of a substance determines what it is. Avicebron argues that the finitude of spiritual substances does not come from matter, but from the form that defines and limits them. Therefore, since all created substances are finite and defined by their form, they must have a structure composed of matter and form, which also includes spiritual substances.

6. THE MATERIALIST FALLACY OF AVICEBRON

In Chapter VI, Saint Thomas refutes Avicebron's position, who, as studied in the previous chapter, held that all substances, both corporeal and spiritual, are composed of matter and form. The Angelic Doctor argues that this view is incorrect for several reasons:

1-The primacy of act over matter: Avicebron suggested that both material and spiritual substances share a kind of "common matter."

Saint Thomas refutes this idea by emphasizing the primacy of act over matter. In his philosophy, matter is considered pure potency, meaning something that has the potential to be but does not have being in itself unless united with a form (which is act). Thus, matter is a lower reality because it does not have its own existence but depends on act to acquire being.

Since spiritual substances (such as angels or souls) are closer to pure act, Saint Thomas argues that they cannot depend on matter, as they are more perfect entities closer to the absolute perfection which is God, who is Pure Act. Therefore, spiritual substances do not need matter to exist, as their nature is eminently actual, not potential. This distinction reinforces the superiority of act over matter in the order of beings.

Avicebron attempts to explain higher realities (spiritual substances) using material principles, which is considered a methodological error. Higher realities, according to the critique, must be understood from formal rather than material principles.

2-Confusion with ancient naturalists: The Angelic Doctor criticizes Avicebron for reverting to the materialist ideas of ancient philosophers, who believed that all reality could be reduced to a single material substance. Avicebron, though he acknowledges the existence of the spiritual, makes a similar error by claiming that all things, both spiritual and corporeal, share a "universal matter." For Saint Thomas, this is unacceptable, as spiritual substances cannot be composed of matter in any sense.

Saint Thomas defends the idea that spiritual substances are radically different from corporeal ones. They do not depend on matter nor are composed by it, as matter is something proper to corporeal things, which require union with a form to exist. Therefore, to claim that the spiritual shares the same matter as the corporeal is, for Aquinas, an unacceptable error, as it implies a reductionist view of reality that does not respect the difference between the spiritual and the material.

3-Generation and corruption: Avicebron also posits that the distinction between species occurs through forms that affect a common matter. But for Saint Thomas, this notion undermines the true understanding of generation and corruption. According to him, prime matter is pure potency, having no act, and therefore cannot be understood as a substance. By reducing everything to matter and denying essential differences between forms, Avicebron eliminates the real distinction between beings.

Avicebron confuses the logical composition of concepts (genus and species) with the real composition of substances.

He argues that species differ from each other due to the forms imposed on a common matter. For him, all the diversity of beings, both material and spiritual, can be explained by a single matter that takes on different forms.

Saint Thomas rejects this view because he considers that it undermines the correct understanding of the processes of generation (when something comes to be) and corruption (when something ceases to be). According to him, prime matter (the most fundamental matter) is pure potency, meaning it has the capacity to be, but in itself has no reality or act until it is united with a form. It cannot be considered a substance or something with its own existence, as it is not actual on its own.

The problem Saint Thomas finds with Avicebron's position is that by reducing all reality to a common matter, the essential differences between beings are diluted. For him, what makes beings different is not just the form they take, but also that prime matter alone cannot account for that diversity. Forms play a fundamental role in making things what they are, and they are not mere modifications of a common matter. By denying these essential differences between forms and reducing everything to matter, Avicebron eliminates the real distinction between beings and the true understanding of how something comes to be or ceases to be.

4-Distinction of potencies: Saint Thomas points out that potencies (capacities or faculties) are distinguished based on their relationship to act, i.e., to what things can become in their full development or perfection. In the case of spiritual substances (such as angels or the human soul), these do not have matter. Therefore, they cannot be related to form in the same way that corporeal or material substances are.

In contrast, Avicebron proposes that all things, both spiritual and material, are composed of a "universal matter." For Saint Thomas, this idea is problematic because it implies an infinite regression of material causes: if everything, including spiritual substances, is based on a common matter, then each thing should have a prior material cause, leading to an endless process without a first cause.

From the philosophical standpoint of the Angelic Doctor, an infinite regression is absurd, as it would prevent reaching a first cause or ultimate foundation of reality. Therefore, he maintains that spiritual substances cannot depend on matter in the same way as corporeal substances, and potencies must be understood in relation to act differently in each case.

What you should remember

1-Primacy of form over matter: For Saint Thomas, higher realities must be understood in terms of their forms, rather than an underlying matter. Matter is viewed as pure potency (the capacity to receive a form) and acquires being only when it has been shaped by a form. Forms, on the other hand, confer identity and being to things, and in spiritual substances, which are not bound to matter, form has absolute priority. This contrasts with Avicebron's position, who considers that all substances, including spiritual ones, share a common "matter."

2-Impossibility of reducing the spiritual to the material: Spiritual substances, such as souls or angels, by their very nature, cannot be composed of matter and form. For Saint Thomas, these substances do not require matter to exist because they are subsistent forms in themselves. Avicebron, however, errs by attempting to unify reality under a universal matter, which inappropriately reduces the spiritual to the material. This is unacceptable because spiritual substances cannot be explained in material terms without losing their inherent nature.

3-Need for formal principles: Saint Thomas argues that to explain diversity and hierarchy within the order of substances, formal principles are needed to determine the nature of each. A single common matter is insufficient to explain differences between things. The plurality of beings must be understood from the diversity of forms, not matter. Forms determine the essential nature of things, while matter is merely a potential principle that is actualized when informed by form.

4-Logical incoherence: According to Saint Thomas, Avicebron's theory involves a series of logical contradictions and denies fundamental principles of metaphysics. By positing a universal matter for all things, it leads to an infinite regression of material causes, which is philosophically incoherent. Additionally, his approach overlooks the principle that higher realities in the order of beings cannot depend on something inferior (such as matter). This creates confusion in the hierarchy of beings and destroys the clear distinction between the spiritual and the material.

7. BEYOND MATTER: THE UNIQUE NATURE OF SPIRITUAL SUBSTANCES

In Chapter VII, St. Thomas delves deeper into his critique of Avicebron's position, arguing in more detail why a common matter for both corporeal and spiritual substances is impossible. The main reason is that matter, by its very nature, is a receptacle for forms and must therefore adapt to the nature of those forms.

1-Distinction between corporeal and spiritual substances: St. Thomas asserts that matter cannot be the same for both types of substances. If corporeal and spiritual substances had a common matter, there would need to be a distinction in that matter before differentiation into their forms, as spiritual substances do not possess quantitative dimensions. This suggests that the distinction does not come from the forms but from the matter itself.

St. Thomas argues that corporeal and spiritual substances cannot share the same matter due to their fundamentally different natures. Corporeal substances have physical dimensions such as length, width, and depth. The matter that makes up these substances is therefore subject to measurement and quantitative division.

In contrast, spiritual substances (such as angels or souls) do not have physical dimensions. They do not occupy space or have physical extension. If it were assumed that matter were the same for both types of substances, it would imply that matter must have a form of distinction capable of accounting for the differences between corporeal and spiritual substances. However, since spiritual substances lack quantitative dimensions, matter cannot be shared between the two categories.

According to St. Thomas, the difference between corporeal and spiritual substances cannot come from the forms because forms are specific to each type of substance (corporeal or spiritual). Instead, the distinction must lie in the matter itself. This means that the matter of spiritual substances must be intrinsically different from the matter of corporeal substances.

2-Reception of forms in matter: St. Thomas explores how matter receives form in both corporeal and spiritual substances and how this demonstrates their differences.

For corporeal substances, matter receives form in a "particularizing" way. This means that corporeal matter receives a specific form that distinguishes it from other forms. This process of reception is limited and weak, as matter cannot fully receive the form due to its own limitations.

Corporeal matter has a capacity for reception that fits its physical dimensions and material characteristics. Form is particularized in matter, meaning that matter individualizes the form according to its specific properties and physical constraints.

Spiritual substances, such as angels and souls, receive form completely and without limitations. This allows them to understand and encompass the form in its entirety, as spiritual substances are not subject to the physical and quantitative restrictions that affect corporeal substances.

Spiritual substances have a capacity for reception that is not limited by physical dimensions. Therefore, they can receive the form fully, meaning they comprehend and assimilate the form completely.

3-Implication for spiritual matter: If spiritual substances can receive form completely, this implies that their matter (if it exists) must be of a very different nature from corporeal matter. Corporeal matter has physical dimensions and is subject to spatial and quantitative limitations. In contrast, spiritual matter, in theory, would not be subject to these physical limitations.

The nature of spiritual matter must be much higher and more sublime compared to corporeal matter. This is because it must be able to receive and encompass form without the constraints imposed by physical dimensions. The ability to receive form in its entirety and without limitations implies that spiritual matter must be intrinsically different and superior to corporeal matter.

4-Relation between potency and act: St. Thomas maintains that higher beings are those closer to act (actualization) than to potency (potentiality). Spiritual substances, being more noble, have matter that is closer to act than to potency. Therefore, spiritual matter is not "pure potency," as in corporeal substances, but rather is already an act, which makes it more perfect.

In terms of matter, potency refers to the capacity of matter to receive and be determined by different forms. Act represents the complete form that actualizes the potentialities of matter. St. Thomas asserts that higher or more perfect beings are closer to act than to potency because perfection is associated with complete realization rather than mere potentiality.

Corporeal matter is "pure potency" in the sense that it is in constant flux and has not achieved full actualization. In contrast, spiritual substances (such as angels and souls) have matter that is closer to act. This means that, although spiritual substances also have form, this form is fully realized and not subject to the limitations and potentialities of corporeal matter. Spiritual matter is not "pure potency" but is already an act, which makes it more perfect.

5-Spiritual matter as substance: If spiritual matter is an act, St. Thomas concludes that there is no real distinction between matter and form in spiritual substances, implying that these substances are simple, not composite. This is because, in spiritual substances, matter and form are one and the same.

Spiritual substances, such as angels or souls, are "spiritual matter." Unlike physical matter, which is passive and receptive, spiritual matter is active and does not have the same limitations.

When it is said that spiritual matter is an "act of being," it means that these spiritual substances are complete and actualized in their existence. They are not in potentiality to become something more; they are already fully realized.

In physical substances, there is a real distinction between matter and form. Matter is the substance that receives form, and form is what determines the essence of matter. However, in spiritual substances, there is no such real distinction. This is because, in the spiritual realm, matter and form are not two separate principles but unify into a single reality.

As a result of the absence of a real distinction between matter and form, spiritual substances are considered "simple." This means they are not composed of material parts or differentiated principles as physical substances are.

In summary, St. Thomas argues that in spiritual substances, matter and form are so unified that these substances do not have a composition or structure that implies separate parts. They are simple and complete in their existence.

6-Precedence of act over potency: St. Thomas reinforces the idea that act precedes potency and form precedes matter. If some forms cannot exist without matter, it is because they are imperfect forms that need matter to exist, which is not the case with spiritual substances. These subsistent forms do not depend on matter for their existence, distinguishing them from corporeal substances.

7-Souls as forms: Even in souls, which are the lowest spiritual substances, St. Thomas explains that while they are united to bodies, they are not composed of matter and form, as a substance composed of matter and form cannot be the form of a body. This is because a form is an act, and matter is pure potency, which would be contradictory.

THE ANGELS

What you should remember

1-Diversity in the reception of forms: Corporeal matter receives forms in a particularized way, that is, not in their entirety. This is due to the nature of corporeal matter, which is closer to potency than to act. In contrast, spiritual substances receive forms in their entirety, which implies a much greater receptive capacity and a nature closer to act.

2-Hierarchy of being: Act is superior to potency in the hierarchy of being. Spiritual substances, being more perfect, must be closer to act and less to potency. Therefore, the matter of spiritual substances must be closer to act than the matter of corporeal substances.

3-Impossibility of common matter: If matter were common to both, it would have to be capable of receiving both corporeal and spiritual forms, which is contradictory. The matter of spiritual substances must be an act or subsistent form, not pure potency.

4-Consequences of avicebron's theory: Avicebron's theory leads to an infinite regression, as each substance would require a prior material cause to exist. It denies the distinction between substance and accidents by reducing everything to a common substance. It undermines the fundamental principles of metaphysics and natural philosophy.

8. RESPONSE TO THE ARGUMENTS OF AVICEBRON

In Chapter VIII, Saint Thomas responds to Avicebron's arguments regarding the diversity and nature of spiritual substances.

1-Diversity in spiritual substances: Avicebron maintains that the diversity among spiritual substances cannot be explained without assuming that they are composed of matter and form. According to him, without this composition, there would be no way for spiritual substances to be distinct from one another.

Saint Thomas refutes this argument by pointing out that diversity in substances does not necessarily require a composition of matter and form. He clarifies that even substances that are pure forms can be diverse due to their acts, as matter alone is not the only principle of diversity. For example, the materials of celestial bodies and elements differ due to the types of acts to which they are ordered, not because of a composition of matter and form. Moreover, Saint Thomas distinguishes between two types of matter: celestial matter, which is in potency for a perfect act and is not subject to change, and elemental matter, which remains in potency to receive different forms and is thus subject to changes and transformations. This distinction illustrates how matter can vary in its characteristics and capabilities according to its degree of perfection.[10]

2-Diversity of forms: Avicebron argues that diversity among forms can only be explained in terms of the matters in which they are present. According to this view, the form itself has no intrinsic diversity, and only matter explains the variety.

Saint Thomas responds that this view is incorrect because the diversity of forms does not depend solely on their matters. Forms can also differ in degree of perfection and imperfection, such that forms closer to matter are less perfect. Therefore, it is not necessary for all forms to depend on matter to explain their diversity; some forms can be distinct due to their own degree of perfection. In this context, Saint Thomas clarifies that the diversity in spiritual forms is not based on matter but on degrees of perfection. Spiritual substances can vary in their degrees of perfection, being more or less perfect depending on their proximity to absolute perfection, which is identified with God in Thomistic philosophy.

3-Spiritual substances and perfection: Avicebron argues that if spiritual substances differ by their degrees of perfection, then they must have matter. This is because, according to him, perfection and imperfection imply a relation of matter and form.

Saint Thomas refutes this idea by explaining that the perfection and imperfection of a substance do not require it to have matter. Perfection and imperfection can be **inherent to the nature of a substance** without it being necessarily material. A spiritual substance can be more or less perfect without needing to be composed of matter and form. Additionally, Saint Thomas argues that perfection and imperfection are not external or accidental properties requiring a material substrate. These qualities represent the **inherent perfection of each thing's nature** and do not depend on matter.

4-Distinction between spiritual substances and God: Avicebron maintains that if spiritual substances lack matter, they could not be distinct from God, who is the very being *(ipsum esse)*. According to this argument, only God would be the being in the absolute sense, and any spiritual substance without matter would be essentially identical to God.

Saint Thomas argues that the absence of matter in spiritual substances does not imply that they are identical to God. Spiritual substances participate in being in a finite manner, while God is the absolute and infinite being. The distinction between spiritual substances and God lies in that God is the subsistent being in itself, whereas spiritual substances merely participate in being.

Saint Thomas clarifies that although spiritual substances lack matter, they are not equivalent to God. God is the absolute and supreme being, while spiritual substances, though immaterial, do not attain this condition of pure and absolute being.

5-Infinity and finitude of substances: Avicebron suggests that if spiritual substances do not have matter, they should be infinite in all respects, like God. According to him, the lack of matter would imply an infinitude that is incompatible with finitude.

Saint Thomas clarifies that spiritual substances, although they lack matter, are not infinite in the absolute sense like God. Spiritual substances have a limited infinitude because they are not subject to matter, but they remain finite in terms of their reception of being from the First Principle according to a determined mode. Absolute infinitude is exclusive to God, the infinitely perfect being, infinite in all respects.

Additionally, Saint Thomas distinguishes between potency in spiritual substances and potency in matter. While potency in matter relates to the capacity to receive forms and material changes, potency in spiritual substances is linked to their own order of being, without relying on a material structure to express their potential.

What you should remember

From Saint Thomas's refutation of Avicebron, the following should be highlighted:

Nature of spiritual substances

-**Diversity without matter**: Spiritual substances can be diverse in degree of perfection without needing an underlying matter.

-**Hierarchical perfection**: There is a scale of perfection among spiritual substances, from the less perfect to the most perfect (God).

- **Direct participation in being**: Spiritual substances participate in being more directly and completely than material substances.

- **Immaterial Nature**: They are simple and perfect entities without the need for underlying matter.

Distinction between spiritual and material substances

- **Spiritual vs. corporeal matter**: The matter of spiritual substances is an act or subsistent form, very different from corporeal matter, which is pure potency.

- **Participation in being**: The manner in which substances participate in being is fundamental to this distinction. Spiritual substances have a more direct and complete participation.

9. THE FIRST CAUSE AND SPIRITUAL SUBSTANCES: A DEFENSE OF CREATION

In Chapter IX, Saint Thomas explores different opinions on the origin and nature of spiritual substances, particularly the position of those who deny that these substances were created. The Angelic Doctor presents and refutes various arguments in favor of this position and then offers an alternative explanation.

1-Spiritual substances without a cause: Some philosophers argued that spiritual substances are uncreated and have no cause. Influenced by the principle *nothing comes from nothing*, they claimed that spiritual substances could not have been created. This principle is used to argue that everything that exists must have a cause, but applying it to spiritual substances led some to conclude that these cannot have an origin because they are not subject to the processes of change and movement typical of material things.

According to this perspective, spiritual substances, such as the soul or angels, do not contain matter and therefore could not participate in the processes of transformation or creation that occur in the physical world. Due to their immaterial nature, these philosophers believed that spiritual substances must exist independently and without an external cause to generate them. Thus, they concluded that spiritual substances have always existed and were not created by any other entity, not even by God.

This error arises from a limited interpretation of creation, which assumes that all change or creation must follow the patterns of the physical world, where material objects require material causes. By failing to understand that spiritual substances can have an origin without depending on matter, these philosophers did not recognize the possibility that immaterial substances could also be created by God, though in a different way from material beings.

2-Hierarchical cause of spiritual substances: Another group admitted that spiritual substances have a cause but argued that not all derive directly from God. According to this view, the higher spiritual substances (such as certain angels or higher intelligences) come directly from God, while the lower spiritual substances (such as human souls or lower-ranking angels) do not have their direct origin in God but derive their existence from the higher substances.

This conception establishes a hierarchical order in which spiritual substances originate in a descending manner. In this scheme, only the highest spiritual entities, those closest to perfection, would be directly connected to God as the first cause. As one descends the hierarchy, the lower substances would receive their being not directly from God but through the higher substances.

This philosophical error arises from the idea that creation must follow a hierarchical model, where less perfect beings depend on more perfect ones. However, this view limits God's creative power by reducing His direct influence only to the highest substances, while the others are relegated to being caused by intermediaries.

The Christian doctrine, on the other hand, teaches that God is the direct cause of all spiritual substances, regardless of their hierarchical level. Each of them, whether higher or lower, has its immediate origin in God, who is the first and direct cause of all being, without intermediaries.

3-Partial creation of spiritual substances: Saint Thomas presents a third erroneous position on the creation of spiritual substances. According to this perspective, the philosophers who defended it admitted that all spiritual substances (such as souls or angels) come directly from God in terms of their being, that is, their basic existence. However, they argued that some essential attributes or faculties of these substances, such as understanding or life, are not received directly from God but derive from higher spiritual substances.

In other words, although they accepted that God is the first cause that gives being to spiritual substances, they claimed that important aspects of their nature, such as the capacity for understanding or life, came from other higher beings in the spiritual hierarchy. In this way, they attributed an intermediate role to higher substances in communicating certain essential aspects to the lower substances.

This philosophical error is based on the idea that not all aspects of a spiritual substance have their immediate origin in God. By holding that important attributes like life or understanding are communicated through intermediaries, they limited God's direct creative action. The correct view, from the perspective of Christian doctrine, is that God is the direct cause not only of being but also of all the faculties and attributes of spiritual substances. God grants both existence and all perfections of each spiritual substance without the mediation of other creatures.

Conclusion

These philosophical errors originate from a misunderstanding of the nature of spiritual substances. Influenced by the material world, many philosophers believed that creation could only occur in the physical realm, where things require a material cause. However, spiritual substances, being subsistent forms, do not have this dependence on matter.

Next, Saint Thomas analyzes the progress in philosophical understanding of the origin of things, from the earliest conceptions to the more advanced ideas of Plato and Aristotle. He argues that the ultimate origin of all being is not the result of a mere physical change but depends on the action of a supreme being, God, who communicates being without the need for mutation or movement. This act of creation is eternal and simultaneous, reaffirming the Christian doctrine of creation.

1-Initial conceptions: Aquinas explains how the early philosophical conceptions of the origin and transformation of bodies were limited. The early philosophers, observing the physical world, only understood that material bodies could originate or change through aggregation (union of parts) and disaggregation (separation of parts) of other bodies. In their view, the change and becoming of things were explained as mere combinations and separations of pre-existing matter.

However, as philosophy advanced, especially with the introduction of the distinction between matter and form, philosophers were able to offer a deeper and more sophisticated explanation of change. According to this new conception, substances do not only change by aggregation or disaggregation of material parts, but change also involves a transmutation or transformation of the forms present in matter.

In this context, matter is seen as the substrate that remains constant throughout change, while form is what transforms, allowing a substance to acquire new qualities or become something different. This advancement in understanding change enabled philosophers to better explain phenomena such as generation and corruption (that is, the birth and destruction of things), beyond mere observable material processes.

2-Becoming according to Plato and Aristotle: Both philosophers recognized that, beyond the observable physical changes (such as the transformation of one thing into another), there exists a higher mode of becoming that involves **participation in being itself**.

According to this view, being is communicated or transmitted to all things by a supreme being. This supreme being does not participate in being as created things do, which depend on others for their existence. Instead, this supreme being is being itself, or *ipsum esse*, meaning its essence is to be, and it does not depend on anything else to exist.

This supreme being, identified as God, is the first cause and the source of all being in the universe. Plato and Aristotle hold that everything that exists receives its being from this fundamental principle, which is not subject to change or movement like material things. God is the origin and foundation of existence, but His influence in the world is not realized through a physical process, but rather as a pure act of being that is communicated to all things.

3-Being without mutation or movement: Unlike the early philosophers, Plato and Aristotle conceive the origin of being in a way distinct from the early thinkers. While early thinkers believed that everything arose through movement or physical mutation (i.e., through visible material changes), Plato and Aristotle recognized that the origin of things can occur without the need for such changes.

In this view, God, as the Supreme Being, does not require movement or mutation to cause being. God does not need to transform, alter, or displace something for it to exist. As the Pure Act of being, or *ipsum esse*, God causes the being of all things without the need for physical change, as He is beyond all movement and mutation.

Following the described criterion, the creation of things would be an eternal and simultaneous act of God. This means that, for Him, the act of causing being would not be subject to time or sequential processes like those we experience in the material world. God would act outside of time, and His creation would be simultaneous, complete, and would not require transformation.

4-The Analogy of light: To illustrate this idea, the Angelic Doctor uses the analogy of light: just as a luminous body illuminates without mutation, God illuminates the being of all things without the need for movement. Similarly, intellectual truths, like premises that generate conclusions, are always true without requiring change.

St. Thomas uses the analogy of light to explain how God grants being to all things without the need for movement or change. This comparison points out that, just as a luminous body illuminates by simply being present without altering its own nature, God causes the being of all things without undergoing any kind of mutation or transformation.

Light simply illuminates by its presence, without changing its own essence. In the same way, God, being the Supreme Being, acts without moving or changing, granting being to creatures in an immediate and constant manner.

Moreover, this analogy extends to the realm of intellectual truths, such as premises in an argument. Premises always generate true conclusions without needing to change or transform. Similarly, the eternal truth of God does not require any alteration to remain valid and effective in the creation of being.

5-Reflection on Plato and Aristotle: Both Plato and Aristotle affirmed that some substances have an eternal existence. For Plato, these substances are the Ideas or Forms that exist in an immutable and eternal realm. Aristotle, on the other hand, also considered the existence of an first unmoved mover or necessary being, which is the ultimate cause of movement and change in the universe.

Despite their belief in the eternity of certain substances, Plato and Aristotle did not ignore the necessity of a fundamental principle that is both immobile and eternal. This principle is the ultimate origin of everything that exists and gives meaning to the order of the universe, even though the substances themselves may be eternal and not created in a material sense.

Christian faith, in contrast, holds that all substances, even the eternal ones according to Plato and Aristotle, were created *ex nihilo* (from nothing) by God. This means that, from the Christian perspective, creation does not only occur in time, but it is an act of absolute creation in which God brings all things into existence from nothing, not merely organizing or modifying preexisting eternal substances.

6-The relationship between divine being and creation: Finally, the Angelic Doctor affirms his position: God does not create out of obligation or internal necessity, but through a free act of His will and wisdom. The divine being is not bound by the same laws that govern the material world, nor is He subject to time. Instead, He grants existence and duration to all things according to His eternal design. This means that creation is neither a forced nor automatic act, but an intentional and wise expression of the divine will, which determines both the nature and the duration of each thing's existence, always from a perspective that transcends time.

What you should remember

Arguments in favor of the uncreated nature of spiritual substances

1-Analogy with matter: It is argued that just as matter has no cause, spiritual substances, being immaterial, would also have no cause.

2-Immutability: It is maintained that spiritual substances, being immutable, could not have been created since creation implies a change in state.

3-Intrinsic perfection: Spiritual substances, as subsistent forms, possess intrinsic perfection that does not require an external cause.

4-Eternity: The idea is invoked that spiritual substances are eternal and therefore could not have had a beginning in time.

Critiques of these arguments

1-Limitations of the analogy with matter: Matter and spiritual substances are ontologically distinct. Matter is pure potentiality, while spiritual substances are acts.

2-Erroneous conception of becoming: Creation is not merely a change of state but a participation in the being of God.

> **3-Ignorance of a higher mode of being**: Those who deny the creation of spiritual substances fail to consider the possibility of a mode of existence that transcends generation and corruption.

The Thomistic position

The chapter under review presents an alternative view based on the philosophy of Aristotle and St. Thomas, according to which:

> **1-God as the First Cause**: All things, including spiritual substances, have their origin in God, who is the First Cause and subsistent being.
>
> **2-Creation without mutation**: Creation does not necessarily imply a change of state in the material sense but a participation in God's being.
>
> **3-Diversity of modes of being**: There are different modes of being, and spiritual substances participate in being in a way distinct from material substances.
>
> **4-Eternity and creation**: The eternity of a substance does not imply that it is uncreated. Many substances can be eternal and simultaneously created by God.

Final conclusions

> -Creation does not necessarily imply a change in time.
> -Spiritual substances, though eternal, can be created by God.
> -The cause of spiritual substances is the creative act of God.
> -Aristotelian-Thomistic philosophy offers a more complete and satisfying explanation of the origin and nature of spiritual substances.

10. GOD AS THE FIRST CAUSE OF ALL SUBSTANCES

In Chapter X, Saint Thomas delves into the discussion about the origin of spiritual substances, refuting the idea that these substances originate from a first created principle and defending the position that all substances, including spiritual ones, have their immediate origin in God.

1-Hierarchy of beings: Saint Thomas examines the question of the hierarchy in creation and the role of God as the First Principle.

Some philosophers, like Avicenna, believe that all things originate from the First Principle, which they identify as God. However, they do not consider that this First Principle acts directly in creation. According to this hierarchical view, the First Principle (God) is so simple and unique that it cannot directly create all things. Instead, creation follows a hierarchical order. The First Principle creates a "first intelligence." This "first intelligence," in turn, generates a "second intelligence." The "second intelligence" produces the soul of the first sphere, and from there, other beings are created, including celestial bodies and other entities. This process continues in a chain of secondary causes that ultimately depend on the First Principle. This model reflects a vision of creation that involves a series of intermediate steps rather than a direct action of God.

Saint Thomas criticizes this idea by arguing that the good of the universe should proceed directly from the intention of the First Agent, who is God. According to his perspective, the order and distinction of the universe should not be merely the result of a necessary order in nature but rather a reflection of the direct intention of God.

Saint Thomas holds that if the universe has order and distinction in its parts, this order must result from God's direct will. In other words, the perfection and order of the universe should not be attributed solely to a chain of secondary causes, but must be a direct manifestation of God's design and intention.

Saint Thomas challenges the view that the First Principle (God) acts through a series of intermediaries to create the universe. Instead, he argues that the good and order of the universe must be the direct result of God's intention, who acts as the First and only Agent, rather than relying on a chain of secondary causes. This emphasizes the idea that the universe is a direct reflection of the divine will, not merely the result of a complex hierarchical process.

2-Production by movement vs. creation: Saint Thomas distinguishes between two modes of production: one that involves movement and another that does not.

In production through movement, things are generated through a process of change. For example, in the generation of plants and animals, things are produced through a series of secondary causes that are interrelated and depend on each other. These secondary causes operate within a framework of causes and effects that trace back to the First Principle, which in this context is God. In other words, the First Principle establishes the initial conditions, but the process of production is carried out through a chain of secondary causes involving movement and change.

In contrast, creation without movement refers to a type of production that does not involve change or movement. This mode of production is considered a pure act of being, attributed exclusively to God. Creation without movement is associated with the creation of immaterial substances and celestial bodies. These beings cannot be produced by secondary causes in the same way as plants and animals, as their existence does not depend on a process of change or movement. Instead, their creation is a direct and absolute act of God, who is the only agent capable of creating such substances.

3-Universal cause vs. particular cause: Saint Thomas also distinguishes between the cause of a nature or form in general (the universal cause) and the cause of a particular form in a specific subject (the particular cause).

Particular cause: It is the agent that produces a form or nature in a specific subject. For example, a horse that begets another horse is the particular cause of the equine nature in that specific horse. In this case, the horse acts as a particular cause because its action produces the equine nature in an individual, but not in the entire species in general. Example: The horse that sires a foal has the ability to transmit the equine form, but only in that particular foal and not in the entire species of horses.

Universal cause: It is the cause that produces the nature or form in general, not just in a specific subject but in all members of a species. The universal cause is the source of the nature itself, in all cases where that nature is found. Example: In the case of equine nature, the universal cause would be the one that produces the nature of horses in general, and not just in a particular horse.

According to Saint Thomas, in the context of creation and production without movement, God is the universal cause that produces the nature of all things.

In production without movement, such as the creation of immaterial entities or creation in an absolute sense, the universal cause (God) is the only one capable of producing the nature itself. This is because production without movement does not involve changes or movements, but a direct and absolute creation of the nature. God, as the universal cause, produces the essence or nature of things without depending on particular beings.

Mathematical concepts, such as numbers or geometric forms, exist in an abstract and universal manner. They do not depend on concrete beings for their existence. Similarly, the universal cause produces the nature of things without depending on particular beings.

4-Virtue of the First Agent: The Angelic Doctor examines the idea of the virtue of the First Agent, which in his philosophy refers to God.

St. Thomas Aquinas asserts that the virtue of the First Agent, God, is absolutely infinite. This means that God's ability to act and produce effects is without limits or restrictions. This infinite virtue contrasts with that of other agents, who may have finite or even infinite capacities in some respects, but never in every sense like God.

The infinity of God's virtue implies that He has the complete capacity to cause anything without relying on any pre-existing material. There are no limitations to His capacity for action.

Other agents, such as created beings, have virtues that can be finite or, in some cases, infinite in certain aspects. For example, a created being may possess significant power, but it is always limited by its nature and existing conditions. Example: A human being has virtues and capacities that are limited by their own nature and the conditions within which they operate.

God, as the First Agent, can produce effects without presupposing any pre-existing potential or matter. In other words, God does not need a prior substance or basis to create something; He can create out of nothing *(creatio ex nihilo)*.

Immaterial substances, which are those that do not have a material basis or cannot be produced by secondary causes or through movement, must be created directly by God. This is because only God's infinite virtue can produce such substances directly and absolutely.

Since immaterial substances cannot result from secondary causes (i.e., processes involving movement or changes in matter), they must be created directly by God. God's infinite virtue allows Him to create these substances without needing an intermediary cause.

What you should remember

Immediate origin in God: Chapter X maintains that all substances, including spiritual ones, come directly from God, refuting the notion that they originate from a created First Principle.

Arguments against Avicenna's position:

> **Lack of divine intention**: If spiritual substances did not proceed directly from God, their existence and order would not be the result of direct divine intention.
>
> **Limitation of divine causality**: Proposing a created first intelligence limits divine causality and introduces an inherent necessity in the order, rather than reflecting free creation.
>
> **Incapacity of secondary causes**: Secondary causes, such as created intelligences, cannot produce absolute being; they can only participate in the divine being.

Arguments for direct creation by God

> **God's universal causality**: God is the universal cause of all things, and His intention encompasses the entire order of the universe.
>
> **Distinction between creation and generation**: Creation is an act of bringing something into existence that did not previously exist, distinct from the process of generation, which involves a change in state.
>
> **Incapacity of secondary causes to create**: Secondary causes only transmit forms, not being itself, which is an exclusive gift of God.
>
> **Need for a First Cause**: Every particular cause requires a universal cause. God is the universal cause of all substances, both material and spiritual.

Conclusion

- Creation is an instantaneous act of God, not a gradual process.
- God is the first and only cause of all substances.
- Spiritual substances do not have an origin independent of God.
- The Aristotelian-Thomistic philosophy defends creation *ex nihilo* (out of nothing).

11. THE UNITY OF PERFECTION IN SPIRITUAL SUBSTANCES: A CRITIQUE OF THE NEOPLATONISTS

In Chapter XI, St. Thomas delves into the nature of spiritual substances and their relationship with the Creator, God. He specifically focuses on refuting the Neoplatonic doctrine that postulates a hierarchy of separate principles from which spiritual substances participate successively.[11]

1-Neoplatonic view on causality: According to the Neoplatonists, God is the immediate cause of the being of all immaterial substances. In other words, they believed that God is the direct creator of everything that exists in the spiritual realm, such as the soul and intellect. This creation act does not imply mutation or movement in the physical sense, as immaterial substances are not subject to these changes. Creation is seen as a pure and direct act of divine will.

The Neoplatonists also held that, although God is the direct source of all spiritual substances, there is an order in how these substances participate in divine goodness. This means that the various spiritual substances not only derive their existence from God, but they can also manifest divine goodness in varying degrees and ways. For example, the soul and intellect may reflect God's goodness at different levels of intensity or perfection. The participation in divine goodness is hierarchically organized, implying that some spiritual beings possess a greater or lesser degree of goodness depending on their proximity to God or their specific nature.

2-Abstract principles and hierarchy: According to the Neoplatonists, there exist abstract principles organized in a hierarchical order. The most general principles, such as the "One" and "being," are the first and most universal. These principles form the highest base of the hierarchy. Following these are the principles of "life," and lastly, the principle of "intellect."

In this view, the first principle (God) is identified with being itself, the most fundamental and universal principle. The next level in the hierarchy is the principle of life, and below this is the principle of intellect. Thus, an immaterial substance that possesses intelligence, life, and being receives these characteristics from different, separate principles. According to this theory, each of these essential qualities (being, life, and intelligence) comes from a distinct abstract principle within the Neoplatonic hierarchy.

Consequently, if there exists any immaterial substance endowed with intelligence, life, and being, it would be an entity by participation in the first principle, which is being itself; it would be living by participation in another separate principle, which is life; and it would be intelligent by participation in a third separate principle, which is understanding. As if we were to say that man is an animal by participation in that separate principle which is the animal, and that he is bipedal by participation in a second principle, which is the biped.

3-Critique of Saint Thomas on Neoplatonic doctrine: Saint Thomas acknowledges that the Neoplatonic doctrine contains some truths, but he considers it incorrect in absolute terms. His critique focuses on the fact that the essential qualities of a substance cannot derive from several distinct principles.

For Saint Thomas, being (or "entity") is the first principle and, therefore, the source of all the perfections that substances can have. By stating that *the effect cannot be simpler than the cause*, he establishes that the complexity and richness of an effect's properties must be reflected in its cause.

He argues that if an effect had multiple causes that were substantially different, the effect itself could not be a simple unity. For example, if being, living, and understanding in an immaterial substance were different causes, then the resulting substance could not be a unique being. This argument is based on Aristotelian logic, which holds that an effect must reflect the nature of its cause.

The properties that are substantially predicated of a thing are intrinsic to its being. In the case of immaterial substances, their being cannot be separated from their living or understanding; that is, for them, living and understanding are modes of being, not accidental additions. Therefore, if something is a being, it must also be living and intelligent from the same source.

Saint Thomas uses the distinction between accidental properties and substantial properties to illustrate his point. Something can be the cause of an accident without being the cause of the substance itself, but in the case of immaterial substances, being, living, and understanding are intrinsically one. This reinforces his idea that *an effect cannot be more complex than its cause*, as the complexity of the cause must be reflected in the effect.

4-Refutation of the separation of principles: According to St. Thomas, if in immaterial substances being, living, and understanding were distinct principles and were added to the essence of the substance as **accidents**, then the Neoplatonic theory would be valid. This means that, in the Neoplatonic view, an immaterial substance would derive its being from one principle, its life from another, and its ability to understand from a third separate principle.

However, St. Thomas maintains that in immaterial substances, living and understanding are not distinct from being. Instead of being added separately, these essential attributes are intimately related to being itself. Therefore, all these aspects (being, life, and intelligence) proceed from the same principle: God.

Now, this doctrine may contain some truth, but considered absolutely, it cannot be true. For when it comes to things that accidentally befall a subject, nothing prevents the first from originating from a more universal cause and what is subsequent from deriving from a lower principle; just as animals and plants partake of the heat and cold of their own elements, while they receive the specific mode of the complexion corresponding to their species from the seminal virtue by which they are generated. For there is no inconvenience in a thing being extended, white, and hot by virtue of different principles.

But when it comes to attributes that are predicated substantially, this is entirely impossible. For all attributes that are predicated substantially of something are intrinsically and absolutely one. And an effect is not reduced to several first principles insofar as they are principles, since the effect cannot be simpler than the cause. This is why Aristotle himself argues against the Platonists: that if the separate principles, "animal" and "biped," were distinct, there would not be an absolutely one bipedal animal.

5-Partial acceptance of the neoplatonic view: While St. Thomas rejects the fundamental Neoplatonic idea that essential qualities (such as being, life, and understanding) come from distinct, separate principles, he is willing to accept certain aspects of the Neoplatonic theory in a more limited context.

St. Thomas acknowledges that the Neoplatonic theory may have some validity in additional aspects that do not affect the essential qualities themselves. For example, he accepts the possibility that certain additional aspects, such as **intelligible species** (which are forms of abstract knowledge or concepts), may derive from an order of higher substances, in accordance with the order established by God.$_{12}$ These **intelligible species** are not essential to the existence of an immaterial substance, but they can be seen as additional elements related to a hierarchical order of realities according to the divine design.

What you should remember

1-Unity of essential perfections: Essential qualities are inseparable and proceed directly from God, not from separate principles.

2-Unity of spiritual substances: St. Thomas critiques the Neoplatonic view that establishes a hierarchy of separate principles, where each perfection of spiritual substances comes from a distinct principle. According to him, this doctrine fragments the unity of spiritual substances.

3-Unique cause: Thomas asserts that all the essential perfections of spiritual substances must derive from a single principle, which is God.

4-Rejection of the neoplatonic analogy: Thomas rejects the Neoplatonic analogy that compares spiritual substances with material compounds, where essential properties may come from different principles. In spiritual substances, essential perfections are intrinsically united and cannot come from multiple principles.

5-Defense of unity and perfection: St. Thomas defends the idea that spiritual substances possess a perfect unity in their being, as all their perfections come directly from God in a single act of creation. This emphasizes the simplicity and perfection of these substances and the omnipotence of God.

12. THE INITIAL INEQUALITY OF SPIRITUAL SUBSTANCES: A CRITIQUE OF ORIGEN

In Chapter XII, Saint Thomas focuses on the discussion regarding the nature of spiritual substances and their beginning. Specifically, he dedicates himself to refuting Origen's position, which claimed that all spiritual substances were created equal by God and that their subsequent diversity was due to their free choices.

1-Critique of initial equality according to Origen: Origen, a 3rd-century theologian, held a particular view on creation and inequality in the universe. According to him, God, being a just author, could not create something unequal without a preexisting diversity that would justify that inequality. This implies that before creation, everything must have been equal to prevent any injustice in the creation of unequal beings.

According to Origen, at the moment of creation, all things were equal. God created a universe in which spiritual substances were all identical. The idea here is that, to maintain justice, all creatures, at the moment of their creation, had to be equal, with no preexisting differences to justify an unequal creation.

Origen believed that inequality among spiritual substances emerged after creation. This inequality was not part of God's initial plan but resulted from the free will of the immaterial substances. According to Origen, these spiritual substances had the freedom to decide and, by using their free will disorderly, deviated from their divine principle. This deviation from God produced differences and inequalities among them.

As a result of this deviation, the need for bodies was created to chain the spiritual substances that had strayed from their original state of equality. These bodies served as a kind of prison for the souls that had departed from the divine order, placing them in a lower and less perfect state.

Saint Thomas criticizes this view for several reasons:

a-Inconsistency: According to Saint Thomas, the idea that all spiritual substances were initially equal and that inequality emerged only later is inconsistent. Spiritual substances are immaterial, and inequality among them cannot be simply explained by material differences as in the case of physical bodies.

b-Perfection and order: Saint Thomas argues that the perfection of the universe cannot be based on absolute equality. Divine perfection and order require diversity and hierarchy in creation. Origen's view denies that order and perfection are achieved through diversity and intended inequality from the beginning.

c-Free will and creation: Origen's theory suggests that the creation of bodies and inequality result from the misuse of the free will of spiritual substances. Saint Thomas argues that this view attributes the perfection of the universe to chance or disordered will, rather than recognizing that divine creation already incorporates an order and hierarchy reflecting God's perfection and justice from the start.

2-Refutation of the concept of multiplicity: Saint Thomas argues that if all spiritual substances were identical and at the same level of perfection, this would indicate a fundamental problem. From his perspective, multiplicity at a specific level of perfection makes no sense in the context of spiritual substances.

In the natural realm, multiplicity within the same species (for example, many individuals of the same animal or plant species) often reflects a form of imperfection or need. For instance, organisms of a species may multiply to ensure their survival or to fulfill specific functions that cannot be attended to by a single individual. Multiplicity here acts as a mechanism to compensate for a deficiency or practical need.

Spiritual substances, in contrast, are considered by Saint Thomas to be more perfect than celestial bodies. In Thomistic thought, spiritual substances have an intrinsic perfection that does not require multiplicity to achieve their purpose. Unlike physical bodies, which may need many individuals to fulfill their roles in nature, spiritual substances do not need to multiply to perform their functions or achieve their objectives.

The idea that spiritual substances would need to multiply to reach their purpose would be a sign of imperfection. Saint Thomas maintains that a perfect spiritual substance should not require the existence of multiple instances of its type to fulfill its function in the universe. The existence of multiple spiritual beings at the same level of perfection would be unnecessary and could indicate that these substances are not entirely perfect in themselves.

3-Perfection and order in creation: Saint Thomas asserts that the perfection of a creature must reflect the perfection of its creator, who is God. God is the perfect being and, as such, creates a universe that must also reflect a form of perfection. This perfection in creation does not manifest in uniform equality among all creatures but in diversity and order.

This means that creatures are not all equal to the same degree but have different levels of perfection and roles within the cosmos. Diversity in creation reflects a deeper and more significant divine order, which organizes the universe so that everything has its place and specific function.

Origen's stance, which proposes absolute equality in creation, is criticized by Saint Thomas. Origen suggests that initially, all things were equal and that inequality emerged later due to the free choices of immaterial substances. This idea, according to Saint Thomas, denies the perfection of the divine order that should be reflected in creation. For Saint Thomas, such uniform equality does not reflect the perfection and order that should characterize the work of a supreme and perfect Being.

Saint Thomas also critiques the idea that the order observed in creation could be attributed to chance. For him, the perfection of the universe cannot be the result of chance, as chance does not provide a reason for being or a purpose. Instead, the order and structure we see in the universe are manifestations of God's intentional and rational design. Attributing order to randomness would be irrational, as it ignores the principle that the universe must reflect divine perfection and order.

4-Justice and perfection in creation: Saint Thomas argues that God created the universe with intrinsic diversity and unequal order. This diversity and inequality are not due to a preexisting variety before creation, but rather are the result of the divine plan to achieve the perfection of the whole. The idea is that diversity and differences in created things contribute to the overall good of the universal order.

According to Saint Thomas, perfection is not found in the uniform equality of all things but in the ordered and diverse structure of the universe. In other words, the perfection of the universe lies in how all parts and beings fit and complement each other to form a harmonious whole. Each thing has its specific place and function within this order, contributing to the general good of the cosmos.

To illustrate this point, he uses analogies from the human body and civil society. In the human body, perfection is found in the cooperation of different organs and limbs that have distinct but complementary functions. Similarly, in a civil society, perfection is achieved through the diversity of roles and responsibilities among its members. In both cases, diversity and differences are not signs of imperfection but are necessary for the functionality and well-being of the whole.

Saint Thomas asserts that God created unequal things from the beginning, not because there was a preexisting diversity in the spiritual realm, but because diversity was necessary for the perfection of the universe. Creation was not based on prior equality but on the idea that the universe, with all its differences and hierarchies, is more perfect as a whole.

Therefore, in the initial creation of things, God produced diverse and unequal things, attending to what is required for the perfection of the universe, and not to any preexisting diversity in things; for this, He awaits the reward of the final judgment, in which He will give to each one according to what they have deserved.

What you should remember

1-Critique of Origen's position: Saint Thomas considers Origen's idea that all spiritual substances were created equal and that subsequent diversity is due to the freedom of choice of these substances as inconsistent. For Thomas, this does not reflect divine perfection.

2-Necessity of initial inequality: Thomas argues that inequality and diversity among spiritual substances are essential for the perfection of the universe. A cosmos composed of identical entities would be imperfect and lack the necessary complexity and harmony.

3-Divine justice: According to Thomas, the creation of unequal spiritual substances is not unjust. True divine justice is manifested in an ordered and diverse universe, where each thing has a specific role in the divine plan.

Key Points

-The initial inequality in spiritual substances is fundamental to the perfection of the universe.
-Origen's position, which implies chance in creation, is unacceptable from the Thomistic perspective.
-Divine justice is reflected in the creation of an ordered and diverse universe, not in uniform equality.

13. KNOWLEDGE AND PROVIDENCE OF SPIRITUAL SUBSTANCES

In Chapter 13, Saint Thomas considers the errors regarding the knowledge and providence of God and spiritual substances—errors that arise from incorrectly comparing these substances to human capacities and operations. Saint Thomas's argument mainly addresses two aspects: the knowledge of singulars (individuals) and the exercise of divine providence over worldly things, especially human actions.

1-Error about the knowledge of singulars: Humans know singulars (concrete individuals) through the senses. For example, we can perceive a particular person, a tree, or a specific object through sight, hearing, etc. However, our understanding does not directly grasp singulars; instead, it abstracts universals from our sensory experiences. Universals are general concepts like "humanity" or "tree" that encompass many particular individuals.

Some thinkers applied this same limitation to God and spiritual substances. Since angels and God do not have bodies or senses, it was thought they could not perceive singulars in the same way humans do. This view argued that spiritual substances could only know universals, meaning general and abstract ideas, because they lack the necessary bodily senses to perceive particular individuals.

Saint Thomas rejects this position because it assumes that God and angels are limited in the same way as humans. According to him, this comparison is erroneous because spiritual substances have a much more perfect capacity for knowledge than humans. God and angels do not depend on the senses to know singulars; their knowledge is direct and encompasses both universals and particular individuals.

In the case of God, His knowledge is absolute and perfect, encompassing everything that exists, including concrete individuals, because He is the cause of all. Saint Thomas argues that limiting God's knowledge and that of the angels to universals underestimates their perfection and fails to recognize the superior nature of their understanding.

Consequently, the Aquinate rejects this view as reductionist, mistakenly assuming that spiritual substances operate under the same limitations as humans, when in reality, their capacity for knowledge is much more perfect and complete.

2-Error about divine knowledge: Some thinkers held that when someone knows something, that knowledge perfects them. Since God is the most perfect being that exists, and nothing outside Him can perfect or surpass Him in nobility, they argued it would be contradictory for God to know something external to Himself, as that would imply created things would perfect Him in some way. According to this logic, God can only know His own essence, as nothing outside Him is worthy of perfecting Him.

Saint Thomas rejects this position for several reasons:

-God is omniscient: To claim that God only knows His own essence implies a limitation on His knowledge, which is incompatible with the idea of divine omniscience. If God is infinite and perfect, His knowledge must also be infinite, meaning He must know everything that exists, both the universal and the particular.

-God as the cause of all: According to Saint Thomas, God knows all created things because He is the cause of everything that exists. Everything that is is known by God in its very essence, as all creatures exist through His creative act. God does not need to be perfected by knowing something external; rather, everything that exists receives its perfection from God. He knows everything exhaustively.

-Divine knowledge does not depend on created things: Unlike human knowledge, which is passive and depends on information received through the senses, divine knowledge is active and perfect in itself. God does not need to "learn" or "acquire" external knowledge; in His divine essence, He knows everything He has created, both universals and singulars, without being affected or perfected by them.

Saint Thomas argues against this idea, as it implies a limitation on divine knowledge that is inconsistent with God's omniscience. God, due to His infinite perfection, does not limit Himself to knowing only His essence, but rather knows all that He has created and can know both the universal and the particular.

3-Error about the necessity of future events: This error posits that if everything is under divine providence, then everything that happens in the world must occur necessarily, leaving no room for contingency or chance. This view is influenced by the Aristotelian concept of causality: each effect has a cause that, once it has occurred or is present, necessarily generates its effect. Under this perspective, if the cause of everything is God and His providence is eternal and perfect, then all future events would be predetermined. This would imply that there is no human freedom or contingent events, because everything is fixed by divine will.

Saint Thomas rejects this conception because it does not do justice to the complete nature of divine providence or to the reality of contingent events. In fact:

-God knows future contingents: Saint Thomas holds that God, in His omniscience, knows all future events, including contingent events—those that could happen in one way or another. This divine knowledge does not impose necessity on those events. God, being eternal and outside of time, knows future events in their entirety but knows them as contingent if they are so, meaning He knows them in their true character, not as necessary.

-Providence does not imply absolute necessity: Although God's providence encompasses everything that occurs in the universe, this does not mean that everything happens by absolute necessity. Saint Thomas introduces the idea of *conditional necessity* and *contingency*. Some events occur necessarily (e.g., the movements of celestial bodies), but others, especially human actions, are contingent and free. Divine providence respects the nature of each thing; thus, it allows contingent events to occur as such without compromising God's omniscience or control over the world.

-Compatibility between providence and freedom: Saint Thomas reconciles divine providence with human freedom and contingency in nature. God knows and guides everything that happens, but He does so in a way that does not eliminate freedom or chance in contingent events. Instead of imposing necessity on everything that occurs, God orders events in a manner that respects the freedom and contingency inherent to their nature.

Saint Thomas responds that, although God's providence encompasses everything, this does not imply that everything happens by absolute necessity. God can know contingent events as such, without diminishing their contingent character. In other words, even though God knows the future, His knowledge does not eliminate freedom or contingency in human actions or in the world.

4-Error about evil and divine providence: Some thinkers, confronted with the existence of evil—especially natural evils (such as disasters or diseases) and moral evils (such as sin and injustice)—argued that divine providence could not fully encompass the lower and corruptible world. This view, influenced by a kind of dualism, maintained that God's providence was focused only on spiritual substances (like angels) and celestial bodies (incorruptible and perfect), while lower things, especially human individuals and particular events, were either beyond its reach or under a secondary providence. This was a way of explaining the presence of evil: God did not directly cause it; He simply did not exercise His providence over certain things that escaped His control or attention.

Saint Thomas vehemently rejects this dualistic conception because it limits God's omnipotence and goodness. He argues that **divine providence**

extends to the entire universe, including both spiritual and material beings, both universals and particulars. There is nothing, no matter how small or insignificant, that is outside the reach of God's providence.

Saint Thomas holds that evil is not a "failure" in God's providence but something permitted within the total order of the universe. God allows evil to exist, not because He desires it or causes it directly, but because a **greater good** may arise from it. In this sense, **evil is a deprivation of good**, but it does not destroy the overall order, as God is capable of integrating evils into His plan to achieve a greater good.

Another important aspect of Saint Thomas's refutation is that evil, particularly moral evil, is a consequence of the freedom granted to rational creatures, such as humans. God respects this freedom, even when humans use it to do evil. However, this does not mean that God's providence is inactive: God continues to govern the universe and is capable of guiding creation toward its ultimate end despite the existence of evil. Thus, evil does not contradict divine providence; rather, it is integrated into it without God causing it directly.

What you should remember

Chapter XIII of the *Tractatus* addresses the errors and arguments surrounding divine knowledge and providence, as well as the capacity of spiritual substances (angels, souls) to know the material world. Errors such as **limiting divine knowledge** to God's own essence, **the inability of spiritual substances** to know particulars, **the denial of divine providence over human acts**, and **determinism** that excludes human freedom are presented and refuted. Saint Thomas argues that God has infinite and perfect knowledge of all things, exercises loving providence over the universe without eliminating human freedom, and that evil is a deprivation of good permitted for a greater good.

Key conclusions

THE ANGELS

> -God perfectly knows everything, both universals and particulars.
> -God's providence encompasses all of creation, including human actions.
> -Human freedom coexists with divine providence.
> -Evil is not created by God, but it is permitted for a greater good.

14. KNOWLEDGE AND PROVIDENCE OF GOD

In Chapters XIV and XV, Saint Thomas explores two fundamental aspects of the divine nature: knowledge and providence. Throughout the text, the profound harmony between God's absolute knowledge and His providential action in creation is revealed, highlighting the perfection and omnipresence of divine wisdom in the governance of the cosmos.

1-Introduction to the defense of divine knowledge: Saint Thomas faces a question that has raised doubts and criticisms among both the general public and philosophers and theologians. The central concern is the possibility that God can know all things, a point that may seem contradictory or difficult to accept at first glance. Saint Thomas seeks to demonstrate that, in fact, God possesses absolute and universal knowledge that encompasses everything that is knowable.

To address this issue, Saint Thomas divides his argument into two main parts:

First, he examines the nature of divine knowledge, defending the idea that God, being the Supreme and Absolute being, has a complete and perfect understanding of all things. He argues that there is no limitation in God's knowledge, and that His understanding includes all aspects of reality without exception.

Then, he addresses divine providence, which is how God orders and directs the universe toward its ends. Saint Thomas maintains that, given God's absolute knowledge, it is also necessary for divine providence to encompass and guide all aspects of creation. Divine providence manifests itself in the perfect planning and execution of the governance of the cosmos.

2-Knowledge of all knowable things: Saint Thomas asserts that in God, substance and understanding are identical. This means that God's being and His capacity to understand are not distinct from one another, but are one and the same. This assertion is based on the idea that God is absolutely simple and that His nature cannot be divided into different parts. If the divine substance (essence) were separated from His understanding, then God would not be the perfect unity that He is, nor the absolute First being.

God's absolute simplicity implies that His being has no parts or internal distinctions. Therefore, His capacity to understand and His being are inseparable and coincide completely. This idea derives from Aristotle's metaphysics, which describes substance as that which exists by itself and completely, without division or multiplicity.

Since God is the absolute and separate Being, His knowledge must reflect this same perfection. As God is the first Being and the principle of all that exists, His understanding must universally encompass all that is knowable. There is nothing that can escape His knowledge, as all that is knowable is included in God's universal reason.

The idea here is that, being the absolute Being, God's knowledge is not limited by the conditions of time or space. God has a total and perfect understanding of everything that can be known, without exception. Thus, no knowable thing can be absent from His knowledge, as His capacity to know corresponds to His nature as the Supreme Being.

3-Similarity to separated forms: Saint Thomas appeals to the idea of separated forms to explain the nature of divine understanding. Separated forms are philosophical concepts that exist in the mind as pure ideals, without depending on the material objects that exemplify them. A classic example is the form of "whiteness" in Aristotelian philosophy. This form represents the idea of whiteness in itself, independent of any actual white object.

If a form like whiteness could exist by itself, it would lack nothing of what its essence implies. In other words, whiteness, as a pure idea, would have all the fullness of what it means to be "white" without needing to refer to a material object that is white. This form would be complete in its essence, as its definition and nature would be fully present in itself.

Saint Thomas uses this similarity to explain how God's understanding functions. Since divine understanding is identical to God's substance, and since God is the absolute and complete Being, His understanding must also be perfect and universal. Just as a separated form would have within itself the fullness of its essence, God's understanding must fully encompass and comprehend all reality that is knowable.

In other words, if we consider divine understanding in terms of its essence, it must possess the fullness of what it means to know. The essence of divine understanding, being completely identified with divine substance, is capable of comprehending all reality without any limitation. Therefore, just as a separated form would be complete in its essence, God's understanding is complete in its capacity to know.

4-Participated and essential knowledge: Saint Thomas introduces the idea that finite beings have knowledge through **participation**. This means that creatures do not know autonomously or completely; rather, their knowledge is derived from something external. In other words, the knowledge that finite beings possess is conditioned by their limited nature and by the way they interact with the world. This knowledge is a participation in the most complete form of knowledge that exists, which is divine knowledge.

On the other hand, God possesses knowledge essentially. This means that divine knowledge is not derived or conditioned by external factors. Instead, God's knowledge is inherent to His own essence. Since God is the absolute Being and the principle of all that exists, His knowledge must be perfect and complete in itself. It does not depend on anything external; rather, it is a manifestation of His own divine nature.

The knowledge of finite creatures, such as humans, is limited by their cognitive capacity and by the way they are made. These beings obtain knowledge by participating in a higher principle or through direct experience of the world. Essentially, their knowledge is fragmented and dependent.

In contrast, the knowledge of God is absolute and total. As the principle of all that exists, God possesses knowledge in its purest and most complete form. All knowledge that manifests in finite beings already exists in God perfectly and completely. The essence of divine knowledge includes all that is knowable, as His nature is the source of all knowledge.

5-Perfect knowledge of the divine substance: Saint Thomas asserts that God perfectly knows His own substance. This means that God has absolute and total knowledge of His own nature. Since God is being in Himself and His essence is identical to His existence, His knowledge of Himself is neither limited nor incomplete. The perfection of His knowledge reflects the perfection of His being. There is no aspect of His own substance that escapes His understanding.

In knowing His own substance, God also knows His own virtue or power. The virtue or power of God refers to His capacity to act and produce effects in the world. By having complete knowledge of His essence, God also knows all the possibilities and capacities that derive from it. His power encompasses all that can possibly exist, whether God produces it directly or through secondary causes.

Divine power is infinite and encompasses all possibilities of existence. This includes everything that can exist in any form, whether created directly by God or by secondary agents. Since God's power is limitless, so too is His knowledge of what is possible within that power. Therefore, God has complete knowledge not only of what exists but also of what may potentially exist.

6-The first substance and its knowledge: In the philosophy of Saint Thomas, God is considered the First cause, that is, the primary and fundamental cause of all that exists. This idea is based on the notion that everything that occurs in the universe has a cause, and if we trace the chain of causes backward, we arrive at a First cause that is not caused by anything else. This First being, which is God, is the principle of all that exists.

As the First cause, God contains within Himself all the principles and realities found in effects. This means that everything that exists in the universe has its foundation in God in an eminent way. In other words, the characteristics, properties, and possibilities of all things are perfectly and totally present in God, although not in a material way, but in an essential and intellectual manner.

Since everything that exists in the universe originates from the First cause, it is necessary that these existences and their characteristics be contained and understood intelligibly in God. God, being the First cause, not only creates the universe but also knows it completely. His knowledge encompasses all details and aspects of things, as all that is real in the world has its root in Him.

God's ability to contain within Himself all the principles of things and His knowledge of all realities and their possibilities confirms that His knowledge is perfect. By knowing the First cause, one knows in a certain way everything that derives from it, as all existence and reality in the universe originate from the essence of God. Therefore, divine knowledge is absolute and complete, encompassing all that can be known.

7-Critique of philosophers and aristotelian interpretation: Saint Thomas critiques the Aristotelian and Platonic interpretation that suggests divine understanding depends on external intelligible forms. According to Aristotle, the divine intellect, which moves the heavens and is responsible for the order of the universe, is related to intelligible forms that are outside of it. In other words, Aristotle seems to imply that God's understanding is based on forms external to Him, rather than being completely autonomous.

Saint Thomas argues that this idea is incompatible with the nature of the divine substance. If God's understanding depended on external forms, this would imply that divine knowledge would not be absolute or independent. Instead, according to Thomas's view, God must have knowledge that does not depend on anything outside of Himself, as He is the absolute being and the first cause of all that exists. This external understanding would call into question the simplicity and perfection of divine knowledge.

The Angelic Doctor maintains that divine understanding does not change or is conditioned by external forms. In his conception, God understands all things according to His own essence, which is infinite and perfect. There is no participation of external forms in divine understanding because God is the supreme and first cause of everything. His knowledge is optimal and absolute, without the need to resort to external objects or to modify His understanding.

This argument underscores that divine understanding is perfect in itself and does not change or evolve from one object to another. Instead of depending on external forms, God understands all things by virtue of His own immutable essence. The supremacy of divine understanding lies in its ability to comprehend all reality without being influenced or conditioned by external factors.

8-Extent of divine providence: Saint Thomas establishes that, just as God has absolute and universal knowledge that encompasses all things, similarly, His providence must also extend to everything that exists. This means that not only does God know all things in their entirety, but He also directs and governs all things toward their respective ends.

Providence is defined as the act of an intelligent being that organizes and directs the universe toward a specific purpose or end. In the case of God, this implies that He not only knows all things but also has a plan or order that directs all things toward their ultimate goal. Divine providence is the mechanism through which God ensures that the universe develops according to His plan and purpose.

Saint Thomas maintains that God is the supreme good and goodness itself. As such, all creation must be subordinate to His will and purpose. This means that all things, from the largest to the smallest, derive their order and purpose from divine providence. The perfection and goodness of God guarantee that the order of the universe is not random but is designed to achieve the best possible ends.

8-The extent of divine providence: Saint Thomas asserts that just as God possesses absolute and universal knowledge that encompasses all things, similarly, His providence must also extend to everything that exists. This means that not only does God know all things in their entirety, but He also directs and governs everything towards its respective end.

Providence is defined as the act of an intelligent being who organizes and directs the universe towards a specific purpose or end. In the case of God, this implies that He not only knows all things, but also has a plan or order that directs all things towards their final objective. Divine providence is the mechanism through which God ensures that the universe unfolds according to His plan and purpose.

Saint Thomas maintains that God is the supreme good and goodness itself. As such, all creation must be subordinated to His will and purpose. This means that all things, from the greatest to the smallest, derive their order and purpose from divine providence. God's perfection and goodness guarantee that the order of the universe is not random, but rather designed to achieve the best possible ends.

The extension of divine providence implies that nothing is beyond the reach of God's governance. Everything that happens in the universe is under the control of His providence and is directed toward the final good that He has determined. Divine providence, therefore, encompasses all aspects of reality, from cosmic events to individual actions, ensuring that everything unfolds according to the divine plan.

9-Providence and the unmoved mover: Saint Thomas uses the concept of the unmoved mover to describe God. This concept comes from Aristotle's philosophy, which argued that in a moving universe, there must be an initial cause that sets everything in motion without itself being moved. For Saint Thomas, God fulfills this role: He is the First Mover who initiates movement and action in the universe without being moved or changed by anything outside of Himself.

In Thomistic thought, God is the principle of all action and movement. This means that everything that happens in the universe has its origin in God's action. As the unmoved mover, God not only initiates movement but also sustains and directs the course of all events. The idea is that without this first cause, the universe could not function or exist as we know it.

In a system of ordered causes, each cause influences the subsequent causes. The preceding cause directly affects the following causes. Since God is the First Cause, His influence is the most supreme and universal. His action is not limited or restricted by prior causes, as He is the source of everything that happens. Divine influence extends to all levels of the universe.

Divine providence is the way in which God directs and organizes all things toward their ends. Saint Thomas holds that God moves all things toward their purposes through His intelligence, which is identical to His being. This implies that God's knowledge and being are not separate but are one reality. Divine intelligence, which is perfect and infinite, guides all actions and events to fulfill the divine plan.

Thus, all things are under divine providence, which is how God ensures that the universe functions according to His plan. Divine providence is a manifestation of the influence of the Unmoved mover, guaranteeing that all movement and action in the universe are directed toward their ends according to divine purpose.

10-The order of the universe and providence: Saint Thomas argues that the order we observe in the universe is not a product of chance or accidental causes but is based on an essential principle. This means that the universe has an organized structure that is not due to coincidences but to an underlying intention and purpose. This essential order is a manifestation of the divine will and plan.

The universe is arranged in a way that reflects the intention of the Unmoved mover, who is God. According to Saint Thomas, God, as the Unmoved mover and First Cause, has designed the universe with a clear and coherent purpose. The organization of the cosmos is not the result of random occurrences but obeys the divine intention for a perfect order.

If God's intention did not extend to all levels of reality, the order of the universe would be incomplete and imperfect. In other words, the universe would be subject to chance and lack coherence. However, since divine providence encompasses everything from the first to the last being, every aspect of the cosmos is organized and directed according to the divine plan. This ensures that the universe is an ordered and perfect system, reflecting God's wisdom and intention.

Divine providence extends to all levels of reality, from the highest to the lowest. This implies that every creature and every aspect of the universe is under the influence and care of God's providence. The organization and purpose of each being are directed toward a divine end, ensuring that everything in the cosmos operates in harmony with God's plan.

11-Perfection of divine providence: Saint Thomas holds that divine providence is perfect in two main aspects: disposition and execution. These aspects guarantee that God's providence is the highest expression of perfection and wisdom.

Perfection in disposition refers to God's ability to plan and order all things perfectly. According to Saint Thomas, divine providence not only establishes a plan for the universe but does so in a way that encompasses all the details of reality. This planning is complete and orderly, ensuring that every aspect of the universe has its place and function within the overall design. Divine providence organizes the cosmos in such a way that it reflects supreme wisdom and purpose.

Perfection in execution refers to how divine plans are implemented in the world. Divine providence manifests in the way that the plans and purposes established by God are carried out. This implies that the implementation of providence is universal and efficient, using various means and instruments to ensure that divine order and purpose are effectively realized in reality. The execution of divine providence is complete and leaves no room for chance, ensuring that the universe operates according to God's plan in a harmonious and effective manner.

12-Angels as executors of providence: Saint Thomas explains that the execution of divine providence does not occur directly by God in all aspects of the universe. Instead, God uses angels to carry out His will and to implement His divine plan. Here are some key points regarding the role of angels according to Saint Thomas:

-**Spiritual Substances**: Angels are considered spiritual substances, meaning non-material beings that exist on a plane distinct from material or bodily beings. This places them in a unique position to act according to divine will, as their spiritual nature allows them to intervene in the world in a way that is not limited by material restrictions.

-**Proximity to the First Cause**: Angels are closer to God, the First Cause, than material beings. This spiritual proximity enables them to have a more direct and clear understanding of divine will. Consequently, they can execute divine providence more effectively and accurately.

-**Messengers and executors of divine will**: Angels act as messengers of God, carrying instructions and performing tasks that correspond to the divine will. They are responsible for implementing divine dispositions and decisions in the universe. This includes guiding human beings, influencing natural events, and coordinating the execution of divine plans in reality.

What you should remember

Divine Knowledge

-**Unity of essence and knowledge**: In God, being and knowledge are one and the same. There is no separation between His essence and His capacity to know.

-**Infinite knowledge**: God knows all things possible and impossible, past, present, and future. His knowledge is perfect and complete.

-**Knowledge by identity**: God knows all things in Himself, as all things exist in Him as their First Cause.

-**Essential vs. participated knowledge**: Divine knowledge is essential, while the knowledge of creatures is participated. God knows fully and perfectly, while creatures know in a limited and dependent manner.

Divine Providence

-**Providence and knowledge**: Divine providence is the manifestation of divine knowledge in the ordering of the universe.

-**Order and purpose**: God has created an ordered universe with a specific purpose, and His providence ensures that this purpose is fulfilled.

-**Unmoved Mover**: God is the First Cause and moves the entire universe toward its ultimate end.

-**Angels as executors**: Angels are instruments of divine providence, carrying out His will in the world.

-Perfection of providence: Divine providence is perfect in its planning and execution, ensuring that the universe unfolds according to the divine plan.

Key points to highlight

> -The unity of essence and knowledge in God.
> -The infinitude and perfection of divine knowledge.
> -The relationship between knowledge and providence.
> -The role of angels in executing divine providence.
> -The perfection and order of the universe as a result of divine providence.

15. RESPONSE TO OBJECTIONS ABOUT THE KNOWLEDGE AND PROVIDENCE OF GOD

In Chapter XVI, St. Thomas responds to several objections raised about the nature of divine knowledge, providence, and the relationship between knowledge and perfection.

1-Response to the first objection: In this text, St. Thomas addresses an objection concerning the capacity of divine understanding and that of angels to know individual particulars. The objection argues that if human understanding cannot know particulars, then neither could divine understanding nor that of angels. St. Thomas refutes this objection with the following key points:

1.1-Reflection of the order of beings: The Aquinate holds that cognitive order must reflect the order and structure of beings in the world. That is, the capacity for knowledge must be in accordance with the level of perfection and universality of the being that knows.

1.2-Universal cognitive virtue: According to St. Thomas, higher beings possess a broader and more universal cognitive virtue than lower beings. This idea is based on the observation that higher bodies in the hierarchy of being, such as celestial bodies, have virtues that encompass a wider range of effects compared to lower bodies.

1.3-Higher knowledge: Similarly, the knowledge of higher beings (like angels) is more universal. This means that these beings not only know the universal but can also know individual particulars. The superior understanding of angels and God allows them to encompass both universal essences and particular cases.

The capacity to know individual particulars is not limited to human understanding. Beings with a more universal and perfect nature, such as angels and God, possess a cognitive virtue that enables them to know both the universal and the singular, in contrast to human understanding, which is restricted by its finite and limited nature.

2-Response to the second objection: In the second objection, it is claimed that understanding is perfected by what it understands. In other words, it suggests that knowledge perfects understanding by receiving and knowing objects or forms. St. Thomas responds to this objection as follows:

2.1-Perfection of understanding: St. Thomas clarifies that the perfection of understanding does not refer to the materiality of what is understood but to the intelligible form, that is, to the abstract representation that understanding has of things. In the case of human understanding, it is perfected by receiving the abstract intelligible forms of sensible objects.

2.2-Divine understanding: Divine understanding is the highest and most perfect, and it does not require any additional form to be perfected. Instead of receiving forms from external objects, divine knowledge is inherent to the divine substance itself. This means that divine understanding is perfect in itself because its knowledge is part of its own essence.

2.3-Intelligible species for angels: The intelligible species (abstract forms of knowledge) that angels possess come from divine understanding. Since divine understanding is the most perfect, the forms it transmits to angels are more perfect than those received by human understanding from sensible things.

Divine understanding does not require additional perfection because its knowledge is already perfectly intrinsic to its own substance. Unlike human understanding, which is perfected through the reception of intelligible forms from sensible objects, divine understanding and that of angels derive their perfection from their connection to the highest and most perfect form: divine knowledge itself.

3-Response to the third objection: In the third objection, it is posed that some events may appear fortuitous or accidental from a human perspective, and the question arises about how these events align with divine providence, which is understood as an ordered and perfect plan.

St. Thomas responds to this objection as follows:

3.1-Perspective of the inferior agent vs. superior agent: St. Thomas distinguishes between the perspective of an inferior agent and that of a superior agent. For an inferior agent, who does not have a complete view of the plan, certain events may seem fortuitous or accidental because they are not perceived as part of a larger design. However, for a superior agent, who has a complete vision and broader knowledge of the plan, those same events can fit perfectly within an ordered plan.

3.2-Example of the trap: He uses an example to illustrate this: if someone sends another person to a place where they know there will be thieves, the assault by the thieves may seem accidental to the person sent, but it is not to the one who planned the act. Similarly, what appears to human knowledge as a fortuitous event may be carefully arranged within the divine plan.

3.3-Divine providence: According to St. Thomas, divine providence is an ordered and perfect cause that encompasses all events in the universe. Although humans may perceive some events as casual or unexpected, these events are actually framed within the eternal plan of divine providence. God, being the superior agent, has foreseen and ordered everything, making even events that seem fortuitous purposeful and placed within the overall plan.

St. Thomas holds that the apparent fortuity of some events from the human perspective does not contradict divine providence. Every event, even if it seems accidental, is integrated within an ordered and perfect divine plan that escapes the limited understanding of inferior agents.

4-Response to the fourth objection: In the fourth objection, the discussion revolves around how causes produce their effects and how this relates to divine providence. St. Thomas addresses this objection by explaining the difference between natural and rational causes and how this relates to God's ability to produce necessary and contingent effects.

4.1-Difference between natural and rational causes: Natural causes, such as physical or biological agents, produce effects according to their intrinsic nature. For example, fire heats due to its nature of warmth, and a seed produces a plant of its specific species because that is its nature. Natural causes act by virtue of the inherent properties of their form and structure.

Rational causes, such as human beings who make decisions and act according to plans and concepts, produce effects based on the conceptual form or idea they have in mind. A rational agent produces an effect according to the intention and conceptual design it has established.

4.2-God's capacity as the First cause: God, as the First cause, has a unique capacity to produce effects of two types: necessary and contingent. The production of necessary effects refers to those that occur inevitably due to the nature of the cause. The production of contingent effects refers to those that may or may not occur, depending on divine will and design.

God's foresight and design allow Him to produce both necessary and contingent effects in the world. God has an eternal and perfect plan that includes both events that must happen by necessity and those that occur contingently. This reflects the perfection of divine providence, which encompasses and coordinates all aspects of the universe.

According to St. Thomas, divine providence not only plans for necessary effects but also orders contingent events in such a way that all are integrated into the universal plan. Although some events may seem contingent from a human perspective, all are framed within God's perfect design.

St. Thomas clarifies that the difference between natural and rational causes is crucial for understanding how effects are produced. While natural causes act according to their nature, rational causes act according to concepts and plans. God, as the First cause, has the capacity to produce both necessary and contingent effects, and all of this aligns with His providence and eternal design. This demonstrates the perfection and depth of the divine plan that encompasses both necessity and contingency in the world.

5-Response to the fifth objection: In the fifth objection, the question of the presence of evil in the world is addressed and how it reconciles with the supreme goodness of God.

5.1-God's goodness and inherently Good effects: God, in His essence, is the supreme good, and all divine action is inherently good. This means that the effects produced by God, in His capacity as the First cause, are good in themselves. God cannot produce evil directly, as His nature is completely good.

5.2-Defects and secondary causes: The defects or evils that appear in some effects are not due to the direct action of God, but to the limitations and deficiencies of secondary causes. These secondary causes are agents that, while acting under divine providence, have their own limitations and can produce imperfect or defective results.

By their own nature and limitations, secondary causes can cause effects that include defects or evils. These evils do not stem from divine intention but from the manner in which these secondary causes operate in the world.

5.3-Permission of evil and greater Good: Divine providence allows for the existence of evils in the world, but this permissiveness is framed within a broader and perfect order. God permits certain evils to occur because, in the context of the overall divine plan, these evils may contribute to the realization of greater goods or necessities for the balance and overall good of the universe.

In the order of the universe, certain evils may be necessary to achieve a greater good that would otherwise not be possible. For example, the presence of suffering or difficulty may lead to the development of virtues such as patience and strength, or may result in the greater good of cosmic and moral balance.

St. Thomas explains that, although God is completely good and produces good effects, the defects and evils in the world are due to the limitations of secondary causes and not to God's direct action. Divine providence permits the existence of certain evils because these may contribute to greater goods or to the overall balance of the universe. This perspective helps reconcile the presence of evil with the supreme goodness of God, showing that even evils have a place in the divine plan to achieve a greater good.

What you should remember

-**Relationship between knowledge and being**: St. Thomas establishes a close relationship between knowledge and being. Superior beings, having a more perfect being, also possess broader and deeper knowledge.

-**Universal and singular knowledge**: Superior beings have more universal knowledge, not because they ignore particulars, but because they can comprehend them within a broader context.

-**Divine knowledge**: Divine knowledge is the most perfect and universal of all. God knows all things, both in their universality and singularity.

-**Divine providence and causality**: Divine providence does not eliminate natural causality but orders it towards a higher end. God allows for contingency and evil to achieve a greater good.

Responses to the Objections

-Limitation of human knowledge: The Aquinate argues that the limitation of human knowledge does not imply a limitation in divine knowledge. Superior beings have a broader cognitive capacity.

-Perfection of divine knowledge: Divine knowledge is not limited to intelligible species but encompasses both the universal and the particular.

-Fortuity and providence: Fortuitous events from a human perspective may be part of a broader divine plan.

-Necessity and contingency: God, as the first cause, can produce both necessary and contingent effects, depending on the nature of secondary causes.

-Evil and divine providence: Evil exists, but it is permitted by God for a greater good.

16. ERRORS OF THE MANICHEANS

In Chapter XVII, Saint Thomas devotes himself to refuting the main doctrines of Manichaeism, an ancient dualist religion that postulated the existence of two opposing cosmic principles: good and evil. From a Christian and Aristotelian perspective, Aquinas argues against this dualistic view, offering a more coherent and rational explanation of the origin and order of the universe.

The Manicheans held three fundamental errors:

1-Cosmological dualism: This is the belief in two opposing and co-eternal principles that explain the origin and structure of the universe. The Manicheans asserted that there were two creative forces or entities governing the cosmos: one dedicated to good and the other to evil. This thought implies that evil has an independent and substantial existence, just like good, which leads to a worldview of constant conflict between these two opposing powers.

In this dualistic view:

> **-The principle of good** is responsible for everything that is light, good, and spiritual.
>
> **-The principle of evil** is the source of darkness, evil, and corrupt matter.

2-Corporeal nature of the principles: They believed that both the principle of good and the principle of evil were corporeal entities, that is, they had a physical and material nature, which is problematic from a philosophical and theological standpoint.

The Manicheans claimed that the creator or author of good was an infinite corporeal light. This means they saw good as something physical and tangible, represented by a light that also possessed intellect, that is, the capacity to think and understand.

Similarly, they asserted that the author of evil was an infinite corporeal darkness. This means that evil also had a physical existence, represented by darkness. This contradicted the Christian view, which understands evil not as a substance or physical entity but as the absence of good, a privation of the perfection due to created beings.

3-Divided governance of the world: The Manicheans claimed that the world was not governed by a single supreme principle, as in the Christian tradition, but by two opposing and conflicting principles: one for good and the other for evil. This idea leads to the following implications:

The principle of good and the principle of evil had equal power over creation. This means that both acted independently, each exerting influence over the world. Thus, good things would be the work of the principle of good, while bad things would result from the action of the principle of evil.

This proposition assumes a duality in the control of the universe, where the forces of good and evil oppose and compete for dominance. It is a concept of divided governance in which there is no single principle ordering the entire cosmos, but rather both rival forces are in a constant struggle to exert their influence.

Refutation of these errors

1-Evil cannot be an active principle: Evil does not have an active existence in itself but is a privation of good or perfection. For something to be active and generate an effect, it must exist in act, that is, in the fullness of being. Since evil is essentially a lack of something that should be present, it has no proper entity or capacity to act.

Saint Thomas uses the relationship between body and soul to illustrate this point. The soul is the act and perfection of the body, that is, what gives it life and existence. Death, considered an evil, is not an active force but the privation of the soul in the body. In this sense, evil produces nothing by itself; it is merely the absence of what should be present (in this case, life).

2-Impossibility of a corporeal intellect: The intellect cannot be something corporeal, since, as Aristotle argues, understanding is a faculty that has the capacity to know all things. If the intellect were something material, it would be limited by the physical characteristics of that matter, which would prevent its universal capacity for knowledge.

If we accept that the creative principle of good has an intellectual nature, as is held by both the Manicheans and philosophical and theological traditions, then it cannot be corporeal. A body is limited by space and time, whereas the intellect, in its immaterial nature, transcends these limitations. Therefore, the Manicheans make an error in attributing a corporeal nature to the principles of good and evil, since the principle of good, being intellectual, cannot be material.

3-Good as the end of all governance: Good is the goal toward which all things tend, as good is what satisfies the natural appetite or desire. All acts of governance or direction, whether in the human or divine realm, seek an end that is considered a good. This is inherent to any kind of order or governance.

In this context, there cannot be a "kingdom of evil" as if evil were an end in itself. Evil, being a privation of good, cannot be a final objective or goal. Therefore, the Manichean idea of two separate kingdoms, one of good and one of evil, lacks foundation, since it makes no sense to speak of a governance that seeks evil as its ultimate end.

Finally, the Angelic Doctor criticizes the mistake of the Manicheans in erroneously applying an observation about the material world to universal principles. The Manicheans noted that, in the physical world, opposite effects (such as heat and cold) arise from contrary causes. Based on this, they concluded that good and evil, being opposites, must also be active principles that generate all things.

However, this reasoning is incorrect because contraries in the material world (such as heat and cold) share a common genus, indicating that they have a superior cause that encompasses them. This common cause is prior and more powerful than the particular causes that produce contrary effects. Thus, contraries are not ultimate or creative principles. Rather, both derive from a single superior principle, which is God, the creator of all.

But it is obvious that they were mistaken in considering the nature of contrariety. For contraries are not completely diverse, but rather agree in something and differ in something else: they agree in genus and are distinguished by specific differences. Therefore, since contraries have their respective opposing causes, by which they are distinguished by specific differences, for the same reason they must have a cause that is common to the entire genus in which they agree. But the common cause is prior and superior to the specific causes, and as one cause is superior, its efficacy is greater and extends to more effects. We conclude, therefore, that contraries are not the first active principles of things, but rather that they all have a first active cause

What you should remember

The main errors of the Manicheans according to Saint Thomas are:

- **Cosmological dualism**: They proposed the existence of two opposing creative principles.

- **Materiality of the principles**: They conceived these principles as corporeal substances, one an infinite light and the other an infinite darkness.

- **Dual governance**: They upheld a dualism in the governance of the universe, with two opposing forces exerting their influence.

Saint Thomas' arguments to refute them:

- **Unity of the creative principle**: Everything that exists comes from one single good and perfect principle. Evil is not a creative force, but a privation of good.

- **Immateriality of the creative principle**: The creative principle must be immaterial and spiritual to be the source of all reality, including intelligence and consciousness.

- **Goodness as the foundation of action**: Everything that acts does so by virtue of its goodness. Evil cannot be an active principle but rather a defect or privation.

- **Unique governance**: The governance of the universe must be singular and in the hands of a good principle. The idea of two opposing principles governing the world is incoherent.

Reasons for the Manichean errors

- **Extrapolating the particular to the universal**: The Manicheans made the mistake of applying to cosmic principles what they observed in the natural world, where opposing forces are often found.

- **Misunderstanding the nature of contrariety**: The Manicheans failed to distinguish between the contrariety of properties and the contrariety of principles.

17. ORIGIN OF SPIRITUAL SUBSTANCES ACCORDING TO THE CATHOLIC FAITH

In Chapter XVIII, Saint Thomas explains the Catholic doctrine on spiritual substances, primarily based on Sacred Scripture and the writings of Pseudo-Dionysius the Areopagite, contrasting it with previously developed philosophical ideas.

1-Origin of spiritual substances: Christian doctrine holds that spiritual substances, such as angels, were directly created by God. This aligns with the belief that everything that exists, both corporeal and spiritual, is the product of God's creative action.

Psalm 148, 2 mentions angels and other celestial powers, affirming that they must praise God. Psalm 148, 5 emphasizes that all things were created by the Word of God, reinforcing the idea that both the physical and spiritual worlds are the work of divine creation.

In his writings, Pseudo-Dionysius the Areopagite (especially in *De Caelesti Hierarchia* and *De Divinis Nominibus*) also emphasizes that spiritual substances, including angels, are the direct product of God's goodness. According to him, God, by His universal goodness, gives existence to the essences of all things, including spiritual substances.

He argues that spiritual substances do not have an independent existence or one generated by other beings; instead, they all come directly from divine goodness, which is the unique source of their being and perfection.

2-Relationship with the divinity: In the second point about the relationship of spiritual substances with the divinity, the text explains how these substances, like angels, are deeply related to divinity according to Pseudo-Dionysius the Areopagite:

Spiritual substances, such as angels, are not created by a multiplicity of divine causes or distinct principles. Instead, all the qualities they possess, such as goodness, being, and life, originate from the unique and absolute goodness of God. This means that there are not several gods or divine principles responsible for different aspects or attributes of these substances.

This point is crucial to refute the view of some ancient philosophers, such as the Platonists, who held that the goodness, being, and other attributes of spiritual substances come from different gods or divine principles. In the Platonic view, the "supreme god" might have been considered the source of goodness, while other gods could be responsible for other aspects of reality. Pseudo-Dionysius rejects this idea by asserting that there is only one God who is the source of all that is good, true, and life-giving.

3-Creation and eternal existence: Christian doctrine holds that spiritual substances, such as angels, have not existed forever. Instead, they were created by God at a specific moment in time. According to the Christian faith, God is the creator of all things, and this includes spiritual substances. Scripture, for example, in Isaiah 40, 26, states that God created all things. Furthermore, Saint Thomas also teaches that everything was created at a determined moment, reaffirming the notion of a unique and specific creative act.

In contrast to the Christian view, Platonic philosophy proposes that spiritual substances, like Ideas or Forms, have an eternal existence. According to Plato, these Forms were not created at a specific moment but have always existed and are eternal. For the Platonists, the Forms or Ideas are permanent and fundamental realities that exist independently of material creation and time.

4-Creation of angels: The question of when angels were created in relation to corporeal creatures has generated various opinions among theologians and the Fathers of the Church.

According to Saint Augustine, the angels were created simultaneously with corporeal creatures. This means that, in his view, both the angels and the physical world were created at the same time by God. For Augustine, God's creative act included the creation of all realities in a single act.

On the other hand, Gregory of Nazianzus and Saint Jerome hold that the angels were created before corporeal creatures. According to this interpretation, the creation of angels preceded the creation of the physical world. This implies that, in the order of creation, the angels existed before God created the material universe.

Another perspective considers the creation of angels in the light, rather than their creation in darkness. This interpretation suggests that the angels were created in a state of purity and clarity, in contrast to the darkness and confusion that could be associated with the state prior to the creation of the world. Light symbolizes divine presence and perfection, while darkness might symbolize emptiness or disorder.

5-Creation and place: The question of place in the context of the creation of angels depends on the sequence in which it is believed they were created in relation to corporeal creatures.

If the angels were created before corporeal creatures, they cannot be assigned a physical place in the conventional sense. This is because place, in philosophy and theology, is associated with the corporeal and material. Angels, being spiritual and non-corporeal substances, would not occupy a physical place in the same way that material creatures do. In this context, physical place would not be relevant to them because, being prior to the creation of the material world, they would not be related to the spatial concepts that apply to physical things.

If the angels were created at the same time as corporeal creatures, then the question of place becomes more relevant. In this case, some theologians have suggested that the angels may have been created in a "luminous heaven." This interpretation implies that, although the angels do not have a physical place in the material sense, they may have been created in a symbolic or metaphorical "space" that reflects their purity and their relationship with divine light. This "luminous heaven" would not be a physical place in the traditional sense, but a spiritual or metaphysical state that aligns with the spiritual and elevated nature of the angels.

What you should remember

-**Divine origin of spiritual substances**: Both Sacred Scripture and the writings of Pseudo-Dionysius the Areopagite clearly affirm that all spiritual substances, including angels, were directly created by God.

-**Immediate creation**: God did not create angels through a causal chain but brought them into existence directly by an act of His will.

-**Angelic hierarchy**: Angels are organized in a hierarchy, with different orders and degrees of perfection, all created directly by God.

-**Non-eternity of spiritual creatures**: Contrary to some pagan philosophies, angels are not eternal; they had a beginning in time.

-**Place of creation**: The question of where angels were created is more complex. Some authors suggest a spiritual place or a "luminous heaven," while others argue that the question is irrelevant for spiritual beings.

18. THE IMMATERIAL NATURE OF ANGELS

In Chapter XIX, Saint Thomas delves into the nature of angels, exploring different opinions on the subject and presenting the Catholic doctrine.

1-Doctrine on angels: Throughout the history of theological thought, there have been views that considered angels as corporeal beings, meaning composed of matter and form. This view is based on interpretations of biblical passages that seem to describe angels with physical characteristics. For example:

> **-In Matthew 18, 10**, it is mentioned that *the angels always behold the face of my Father who is in heaven*, which could be interpreted as an indication that angels are physically located in heaven.
>
> **-In Isaiah 6, 6**, a celestial being, a seraph, is described with wings, which also suggests a physical image.
>
> **-In Daniel 10, 5**, an angel is mentioned with a detailed bodily appearance.

The bodily representations in Scripture have led some to think that angels are not purely spiritual but possess some form of physical existence. This interpretation is influenced by the tendency to understand the spiritual through physical analogies, which are more comprehensible to human beings.

Despite these interpretations, traditional Christian doctrine holds that angels are incorporeal and immaterial beings. This belief is based on several theological and scriptural reasons:

-Biblical terminology: Scripture describes angels as "spirits." For example, in Psalm 103, 4, it is said: *Who makes his angels spirits*, and in Hebrews 1, 14, they are described as *ministering spirits*. The term "spirit" in the Bible is generally associated with the incorporeal, that is, with what does not have a physical or material existence.

-Concept of virtue: In the Psalms and the Gospel of Saint Luke, angels are described as "virtues." A virtue is a quality that is not material; it is considered a property of immaterial entities. This suggests that angels are not composed of matter and form but have a purely spiritual existence.

-Focus on the immaterial: Christian doctrine asserts that angels, being spiritual beings, do not have a material or physical existence. This means that, although they may appear in forms comprehensible to humans in Scripture, their essence is entirely spiritual and immaterial.

2-Authority of Tradition

2.1-Pseudo-Dionysius the Areopagite: In his works, such as *De Divinis Nominibus* and *De Caelesti Hierarchia*, Pseudo-Dionysius the Areopagite asserts that angels are incorporeal and immaterial. For him, angels do not have a physical existence; they are essentially spirits or purely spiritual entities.

Pseudo-Dionysius acknowledges that our capacity to understand and conceive the spiritual is limited. Because of this, he uses material figures and representations to illustrate the nature and hierarchy of angels. Although angels themselves are not material, physical images serve as aids for the human mind to grasp their spiritual reality. For example, physical descriptions or symbolic figures can be used to represent how angels are organized and how they interact in the spiritual realm.

Pseudo-Dionysius describes angels as "heavenly intellects" or "divine minds," which reinforces their incorporeal nature. These terms suggest that angels are forms of pure intelligence, with no material foundation, operating on a completely spiritual level.

2.2-Saint Augustine: He also considers angels to be spiritual and intellectual beings. According to his perspective, angels do not have material corporality but are purely spiritual entities that operate at the intellectual level.

While Saint Augustine holds that angels are incorporeal in relation to humans, he also recognizes a distinction in comparison to God. According to him, although angels are spiritual and do not have a physical body, in comparison to the pure spirituality of God, angels might still seem material. This implies that, from a divine perspective, even the incorporeality of angels is relative and does not compare with the absolute simplicity and purity of God.

2.3-Saint Basil: In the *Hexaemeron*, he describes angels as being *in the light and spiritual joy*. This means that their location should not be understood in physical terms, but in terms of the spiritual experience and joy they are in. Angels exist in an elevated spiritual state and in divine light, which is a way of describing their connection to spiritual reality and their participation in divine happiness and fullness.

2.4-Saint Gregory of Nyssa: In *De Homine*, he affirms that intelligible beings, such as angels, are found in *intelligible places*. This suggests that angels exist in a type of place that corresponds with their intellectual and spiritual nature, rather than a physical place. These "places" are understood in terms of their relation to intelligence and spiritual reality, not in terms of physical space.

3-Representations and bodily forms: In Scripture, angels are sometimes described with bodily forms or physical figures. These representations must be understood allegorically, meaning they should not be taken literally. The physical descriptions of angels do not indicate that they actually have material bodies. Instead, they serve as visual aids to facilitate the understanding of spiritual realities that, by their very nature, are immaterial and difficult for the human mind to grasp. Bodily figures, therefore, act as metaphors or symbols that allow human beings to conceptualize and relate to the spiritual nature of angels.

The terms that describe human affections, such as anger or concupiscence, when applied to angels, must also be understood symbolically. Angels do not experience human passions or emotions in the same way as human beings. Instead of having actual feelings like anger or concupiscence, these terms are used to reflect the way angels act in conformity with divine law and their roles in the spiritual realm. In other words, when it is said that an angel acts with "anger" or "concupiscence," it is describing how the angel carries out its functions in relation to divine actions, not that the angel truly experiences those human emotions.

4-Presence and movement of angels: Angels, being incorporeal and immaterial beings, do not occupy physical space as material bodies do. Instead of being located in a physical place, their "presence" is spiritual. This means that angels relate to places not through physical location, but through a spiritual or virtual connection. The presence of angels in a place refers to their relation to that place, such as their influence or their function within it, rather than a tangible physical location. This view is supported by theologians like Pseudo-Dionysius, who explains that when it is said that angels are in a place, it refers to their relation to the place, not their physical presence within it.

Similarly, the movement of angels should not be understood in physical or bodily terms. Angels do not move in the same way that material bodies move through space. Rather, their "movement" is symbolic of their spiritual activity and influence. When the movement of angels is described, it refers to their spiritual activity, such as acting in different areas or fulfilling various functions in the spiritual realm, rather than a physical displacement. This type of movement reflects how angels exert their influence and fulfill their roles in the spiritual plane.

What you must remember

- **Authority of Scripture**: St. Thomas cites numerous biblical passages to support the idea that angels are incorporeal spirits.

- **Authority of Tradition**: He relies on the teachings of authors like Pseudo-Dionysius the Areopagite, St. Augustine, St. Basil, and St. Gregory of Nyssa to ground his position.

- **Allegorical interpretation**: Biblical descriptions that seem to attribute bodily characteristics to angels are interpreted allegorically to convey spiritual truths.

- **Philosophical reasoning**: St. Thomas uses philosophical arguments to demonstrate the incompatibility of corporeality with the spiritual nature of angels.

19. THE ORIGIN OF EVIL IN ANGELS: A THEOLOGICAL DEBATE

In Chapter XX of his work, Saint Thomas Aquinas studies the distinction between angelic spirits, particularly the difference between good and evil spirits. The Angelic Doctor examines various opinions about the nature of demons and the reasons why some angels became evil. This is the final chapter of the *Tractatus*, which was left unfinished.

1-Differentiation between good and evil spirits: Saint Thomas begins by affirming that both sacred doctrine (the teaching of the Church) and Sacred Scripture (the Bible) make a clear distinction between these two types of spirits.

Good spirits: They are described as "ministering spirits." This means that they serve God and assist human beings destined for salvation. Aquinas cites Hebrews 1, 14 to support this claim, where it is mentioned that these spirits are sent to serve those who will inherit salvation, i.e., the faithful and the righteous.

Evil spirits: On the other hand, evil spirits are described as "unclean spirits." These spirits seek rest or repose but join with others worse than themselves, as mentioned in Matthew 12, 43. In this passage, Jesus speaks of an unclean spirit that leaves a person, seeks a place of rest, and, finding none, returns to discover the house (the person) empty and in worse condition, bringing with it other, more wicked spirits.

According to Christian doctrine, good spirits fulfill a divine mission by helping the faithful, while evil spirits seek to cause harm and associate with other malevolent spirits to increase their negative influence.

2-Errors concerning the nature of demons: Saint Thomas addresses and refutes several erroneous theories about the nature of demons. His goal is to correct misunderstandings about why demons are considered evil and what their true nature is from a theological and philosophical perspective.

2.1-Evil does not exist by itself: Saint Thomas holds that evil does not have its own existence; it is not an entity that exists independently. Instead, evil is a privation of good, something that occurs when something good is absent or altered. Therefore, demons, like any other evil thing, cannot be evil by nature, as evil has no independent existence but is a corruption of good existence.

2.2-Demons do not corrupt everything or themselves: If demons were inherently evil, they would be capable of corrupting all things, including themselves. However, this does not occur in reality, as there is no evidence that demons have the ability to corrupt everything they touch or even to totally corrupt themselves.

2.3-Creation by a good God: According to Christian doctrine, God is good and creates only good things. If demons were inherently evil by nature, this would imply that a good God created something evil, which would be contradictory. Since God only produces what is good in itself, the existence of demons cannot be due to the creation of something inherently evil.

2.4-Stability in evil does not indicate an evil nature: Persistence in evil is not a characteristic that defines the nature of evil but is rather a property of good. Stability in good is what demonstrates virtue and perfection. Therefore, the persistence of demons in evil does not prove that their nature is evil.

2.5-Aspects of good in demons: Although demons are evil entities, they still retain aspects of good. For example, they possess existence, life, understanding, and desire, all of which are positive aspects. These attributes show that, while demons act in a wicked manner, they are not inherently evil in their entirety, as they retain elements of goodness.

3-Demons with natural inclinations toward evil: Saint Thomas addresses the idea that demons, though not inherently evil, could have a natural inclination toward evil.

3.1-Intellectual nature and natural tendencies: Saint Thomas argues that demons, if considered as purely intellectual beings (i.e., entities without bodies), cannot have natural inclinations toward evil. The reason is that intellectual nature is oriented toward good. In Thomistic philosophy, intelligence or reason seeks the good as it is the natural object of its knowledge and desire. Therefore, if demons were purely intellectual, their natural tendency would be toward good, not evil.

3.2-Natural inclinations and corporeal beings: Natural inclinations toward evil might be characteristics of corporeal beings, those who have a body and are subject to passions and material desires. In corporeal beings, inclinations toward evil can arise due to the weaknesses and limitations inherent in physical existence and passions. However, demons, being incorporeal beings, would not have these natural inclinations toward evil derived from physical existence.

In other words, the natural inclinations that could lead a being to act in an evil way are typical of corporeal life, where physical passions and desires can influence behavior. Incorporeal beings, such as demons, do not have these physical influences and, therefore, should not have natural inclinations toward evil due to their intellectual nature.

3.3-Rejection of natural inclination to evil: Therefore, if demons were purely intellectual, their inclination toward evil could not be a natural characteristic. The inclination toward evil, rather than being an inherent tendency in their nature, should be understood as a deliberate choice or a corruption of their free will. In Saint Thomas' view, demons choose to do evil rather than having a natural inclination toward it.

4-Demons as corporeal beings: The text addresses the question of the nature of demons and the different opinions on whether demons have a corporeal nature or if they are fallen spiritual beings.

4.1-Demons as airy bodies: There is a theory that demons are corporeal beings, specifically airy bodies. This implies that they have a physical nature, although not necessarily like human or animal bodies, but rather in a more ethereal or subtle form. According to this view, by having a corporeal and sensitive nature, demons would experience passions and inclinations toward evil, similar to living beings with physical bodies. This idea suggests that their inclinations toward evil might stem from their physical nature and the passions associated with it.

4.2-Rejection of the corporeal view: However, not everyone accepts this view. Most theologians and philosophers, including some Doctors of the Church, reject the idea that demons are corporeal beings. Instead, they hold that demons are spiritual beings who were originally created by God as angels.

4.3-Fallen angels: This alternative view, upheld by some Doctors of the Church and also by Neoplatonic thinkers, asserts that demons could have been angels who fell from grace due to their free use of will. According to this perspective, angels were created as pure spiritual beings without a corporeal nature. When some of these angels used their free will to reject God, they became demons. In this case, the evil in demons would not arise from a natural or corporeal inclination but from a free and conscious decision to rebel against God.

4.4-Distinction between views: The key distinction here is that demons are not considered corporeal beings by most theologians, who prefer the explanation that they are fallen angels. This implies that the characteristics and behaviors of demons are not due to a physical nature but to a moral and spiritual fall from an originally good condition.

5-Evil as a result of freedom: According to St. Thomas, evil in demons does not arise from an inherently evil nature but from a free choice they made. In other words, demons became evil due to their own free decisions, not because they were created with a natural inclination toward evil.

Demons were originally good. This is based on the idea that God, in creating the angels (some of whom would become demons), created them as good beings. Goodness is an essential characteristic of divine creation.

Despite their initial state of goodness, demons deviated and became evil due to their free decisions. According to St. Thomas, demons exercised their free will to reject God and His grace, which led them to a state of wickedness.

Pride, or excessive self-esteem, is a key factor in this deviation. The demons, in choosing to oppose God's will, demonstrated a rejection of divine authority and a self-sufficiency that resulted in their fall. This pride was, therefore, the catalyst for their transformation into evil beings.

The notion that evil results from free choice emphasizes the importance of free will in morality. The demons, having the capacity to choose freely, decided to rebel against God. This concept also implies that evil is not a created attribute nor inherent to the nature of creatures, but rather the result of moral and free decisions that lead to corruption and wickedness.

6-Scriptural evidence: St. Thomas supports his argument regarding the fall of the angels and their transformation into demons using Scripture.

6.1-Isaiah and Ezekiel: These two Old Testament prophets are used to show how the fallen angels became evil through their own transgression. Specifically:

In Isaiah 14, 12-15, there is a reference to the fall of a being who aspires to rise above God and is punished for his pride. This passage is traditionally interpreted as describing the fall of Lucifer, the angel who rebelled against God and became Satan.

In Ezekiel 28, 12-17, there is mention of a being who was perfect in beauty and wisdom but became corrupt due to pride and became the subject of judgment. This passage is also understood as an allusion to the fall of an angel, linking transgression with corruption and divine judgment.

By referring to these biblical texts, St. Thomas provides scriptural support for his doctrine that the fall and evil of demons come from their free choice to disobey God. This also demonstrates that the teaching on the fall of the angels and the nature of evil has a basis in Scripture, reinforcing the validity of his theological argument.

7-Platonic view and other opinions: The Platonic view on the transformation of human souls into demons is also addressed.

This view is based on Platonic philosophy and the ideas of Plotinus, a Neoplatonic philosopher. According to this perspective, human souls could transform into demons based on their actions and deeds during their existence. In other words, the character and actions of a person in life could determine their fate in the afterlife, including the possibility of becoming a demonic being.

St. Thomas rejects this idea. According to his theological and philosophical perspective, human souls do not transform into demons. He argues that demons are fallen angels, beings created as pure spirits by God. Human souls, in contrast, are different in nature and substance. They do not have the capacity to become demons due to their distinct nature.

Demons became evil through their free choice and rebellion against God, not through a transformation of human souls. Human souls have their own destiny and do not transform into demons as a result of their actions in earthly life.

THE ANGELS

St. Thomas emphasizes the essential difference between humans and angels. Humans, having a different nature and being part of a distinct created order, cannot become demons, who are entities of a higher spiritual order.

What you should remember

> **-Existence of good and bad angels**: The existence of good and bad angels is a fact accepted by the Church, based on Scripture and Tradition.
>
> **-Causes of evil in angels**: Several causes are explored: 1- Some, following the Manichaeans, argued that demons were evil by nature. 2- Others suggested that demons had a natural inclination toward evil, though they were not evil by nature.
>
> **-Free fall**: Most theologians, including St. Thomas, maintained that angels became evil through a free act of will.
>
> **-The nature of angels**: They are incorporeal.
>
> **-The role of free will**: Angels, as intelligent and free beings, can choose between good and evil.

BY WAY OF AN EPILOGUE

1-What is the main purpose of the *Tractatus de Substantiis Separatis*?
The main purpose is to critically analyze various philosophical theories about the nature and existence of angels and to present the Catholic doctrine, providing a solid basis for the philosophical and theological understanding of angels.

2-What topics are covered in the first part of the work, and how is it structured?
The first part, philosophical in nature, is divided into 17 chapters that examine and refute the theories of ancient philosophers on immaterial substances, including Plato, Aristotle, Avicebron, and Avicenna.

3-Who did Saint Thomas dedicate the *Tractatus* to?
Saint Thomas dedicated the *Tractatus de Substantiis Separatis* to his friend and collaborator Friar Reginald of Piperno, who also took on the task of editing the incomplete works of the Angelic Doctor after his death in 1274. The authenticity of the treatise has never been questioned.

4-Why is the *Tractatus* considered an important work within the Thomistic corpus?
It is a key work due to its historical-philosophical analysis, its synthesis of the Catholic doctrine on angels, its contribution to the study of rational psychology, and for illustrating Saint Thomas's philosophical method.

5-Why is the philosophical aspect considered predominant in this work?
The work remains incomplete, and although it includes a theological section, this was not fully developed. Hence, the philosophical part, focusing on the analysis of opinions on immaterial substances, predominates.

6-What is Saint Thomas's stance on the origin and nature of angels in the second part of the work?

Thomas holds that angels, or spiritual substances, were directly created by God, are incorporeal and immaterial. However, this theological section remains unfinished.

7-What criticisms did the *Tractatus* receive, and how do they relate to Thomas Aquinas's theses?
Medieval philosophers, especially Franciscans, criticized the nature of immaterial substances, their relationship with God, and their knowledge of the sensible world, which sparked debates within scholasticism.

8-How is this work viewed in terms of importance according to different authors?
Although classified as an *opusculum* (minor work), Étienne Gilson considers it historically rich, Joseph Eschmann values it as a key metaphysical text, and Karl Henle describes it as a refined synthesis of Platonic doctrine.

9-What is known about the date of composition of the *Tractatus*, and what opinions exist on this matter?
The exact date is uncertain. Mandonnet and Glorieux suggested placing its composition between 1272-1273. Walz proposed a longer period, specifically 1261-1269. Callus dates the work to 1272. Meanwhile, Grabmann uses May 18, 1268, as the *terminus a quo*, meaning the earliest possible time when the book could have been written or finished. Alternatively, the year 1270 has been proposed as another possible *terminus a quo*. However, Vansteenkiste disagrees with this date and prefers placing it earlier, around 1259.

10-Why is the composition date of the *Tractatus* important?
Knowing the approximate date helps to understand the historical and philosophical context of the work, as well as its relation to other writings by the Angelic Doctor and contemporary debates.

11-What structure does Saint Thomas propose in his investigation of angels in the Prologue of the *Tractatus de Substantiis Separatis*?

Saint Thomas divides his investigation into two stages: the first is dedicated to the conjectures of antiquity and how they can help in understanding the Christian mystery; the second focuses on presenting the Church's teachings about angels.

12-How does Saint Thomas justify the relationship between devotion and study in the Prologue of his work?

Saint Thomas acknowledges the importance of prayer but justifies his intellectual approach as a complement to devotion. He suggests that the study of angels is a form of prayer and a way to make up for time not dedicated to the office of praise.

13-What criterion does Saint Thomas establish for evaluating ancient conceptions of angels?

Saint Thomas proposes accepting what in the ancient conceptions aligns with faith and rejecting what contradicts Catholic doctrine, in an effort to integrate ancient knowledge with Christian teaching.

14-What characterizes the so-called "naturalist" philosophers?

The naturalist philosophers of ancient Greece sought to explain the origin and principles of the universe based on natural elements and processes, without resorting to mythological or supernatural explanations. They focused on *physis* (nature) and are considered precursors to scientific thought.

15-What did Thales of Miletus consider to be the fundamental principle of all things?

Thales of Miletus believed that water was the basic and fundamental principle of all things. He thought that everything originated from water and that it had the capacity to transform into other elements.

16-How did Heraclitus explain the nature of reality, and what was his fundamental principle?

Heraclitus argued that change and transformation are essential characteristics of reality, and his fundamental principle was fire, which symbolized the dynamism and constant process of change in the universe.

17-What was Empedocles' contribution to the understanding of the fundamental principles of reality?

Empedocles proposed that the fundamental principles were four elements: water, air, fire, and earth. These elements combined and separated through two opposing forces, Love and Discord, which explained the variety and complexity of the world.

18-How did Anaxagoras's view differ from that of other philosophers regarding fundamental principles?

Anaxagoras introduced the idea of an incorporeal principle called *nous* (understanding or mind), which organized and shaped matter. Unlike other philosophers who focused on material principles, Anaxagoras added this non-material principle that separated and ordered the parts of the cosmos, though he did not consider it the ultimate foundation of reality.

19-What was Plato's approach to overcoming the limitations of the naturalist philosophers?

Plato proposed that the senses do not provide true knowledge due to the constant change of material bodies, and he postulated the existence of entities separate from matter that represent fixed truths, which the soul can access to know the truth.

20-What problem did the ancient philosophers of nature see in the knowledge of corporeal things?

They held that humans could not know the truth with certainty because of the continuous flow of corporeal things and the deception of the senses.

21-How does Plato classify entities that are separate from matter?

He classifies them into two main types: mathematical entities (numbers and figures) and universal entities (Ideas or species).

22-What is the nature of mathematical entities according to Plato?
Mathematical entities, such as numbers and geometric shapes, do not exist in the physical world but in the World of Ideas. The triangles and circles we see are imperfect manifestations of perfect Ideas or Forms.

23-What are universal ideas according to Plato?
They are universal concepts such as justice or beauty that do not exist materially, but just or beautiful things in the world partake in these Ideas.

24-What is the difference between mathematical entities and universal Ideas?
Mathematical entities can appear in multiple instances, whereas universal Ideas are unique and represent the essence of things.

25-How does Plato organize knowledge and reality?
Plato organizes knowledge and reality in hierarchical levels, where mathematics represents knowledge closer to the sensible world, while universal Ideas, such as the Good, are at the top.

26-What place do the One and the Good occupy in Plato's hierarchy?
The One is the supreme principle of unity, and the Good is the divine principle that gives meaning and value to all other Ideas. Both are at the peak of Plato's hierarchy.

27-What is the role of secondary gods in Plato's philosophy?
The secondary gods are divine entities that participate in the One and the Good in a limited way, and they have roles in the cosmic order, acting as mediators between the sensible world and the One.

28-How does Plato explain the movement of bodies in the cosmos?
Plato argues that bodies do not move by themselves but need a principle of movement, which is the soul. The movement of celestial bodies and everything in the cosmos depends on the influence of the soul.

29-What does Saint Thomas reject regarding universals and mathematical entities?

Saint Thomas rejects the Platonic idea that universals exist separately from particular objects. Following Aristotle, he states that universals or essences (such as humanity or redness) do not exist outside of the individuals that manifest them but are characteristics of particular things.

30-What method does Aristotle propose for investigating separate substances?

Aristotle proposes a method based on the study of movement, observing the sensible world and relying on reason and experience, to investigate immaterial substances, such as the soul or God.

31-What principle does Aristotle establish regarding movement?

Aristotle establishes that *everything that moves is moved by another*, as there cannot be an infinite chain of movers. For movement to exist, there must be a first unmoved mover that is not moved by anything else.

32-Why must the first mover be incorporeal according to Aristotle?

The first mover must be incorporeal because no finite force can move eternally. A material body or limited force cannot generate continuous and infinite movement due to its natural limitations, so the first mover must be an infinite entity, not subject to physical constraints.

33-What relationship does Aristotle establish between the good and movement?

Aristotle holds that the good acts as an unmoved mover in the sense that it moves desirous beings toward it without directly participating in movement. The good is a final cause that directs beings toward their perfection and purpose, without being the primary mover of motion.

34-What does Aristotle assert about the eternal movement of celestial bodies?

Aristotle asserts that not only the "first mover" (the first heaven) moves eternally, but also the lower spheres, meaning all celestial bodies. Each of these bodies has its own soul, which means they are animated and have an "appetible object," an end toward which their movements tend.

35-How does Aristotle attempt to determine the number of immaterial substances?

Aristotle tries to determine the number of immaterial substances based on the number of celestial movements. Avicenna (an interpreter of Aristotle) offers a different version, counting not the movements but the planets and upper spheres, which generates a debate about the correspondence between celestial movements and intellectual substances.

36-How is the cosmos hierarchically organized according to Aristotle?

In the Aristotelian system, the cosmos is organized in a hierarchical structure where each element has a specific role and is subordinated to a higher level. This hierarchy is based on the existence of a supreme heaven or first heaven, which acts as the first moved mover responsible for the motion of all other heavens.

37-What difference does Aristotle establish between the sublunary world and the celestial world?

In the sublunary world, the only animated bodies are those of animals and plants, while in the celestial world, the celestial bodies possess souls that animate them. Aristotle does not admit that simple elements, such as air or fire, are animated, as they lack the proper organs for this.

38-How does Aristotle explain the dual order of intellectual substances?

Aristotle proposes that between humans and the supreme god, there are two types of intellectual substances: separate substances, which are the ends of celestial movements, and the souls of the spheres, which move the celestial bodies through desire and appetite.

THE ANGELS

39-What does Thomas Aquinas criticize about Aristotle's view of the relationship between immaterial substances and celestial movements?

Thomas Aquinas criticizes Aristotle for limiting the number of immaterial substances based on the number of celestial movements. He argues that this is unreasonable, as higher beings should not be subordinated to lower ones, and immaterial substances have a greatness that surpasses celestial bodies.

40-What distinction does Thomas Aquinas introduce regarding the ends of celestial movements?

Thomas Aquinas introduces the distinction between proximate and remote ends, suggesting that the proximate end of the movement of the first heaven does not have to be the supreme immaterial substance (God), but that other orders of intermediate substances may exist between celestial bodies and God.

41-What does Avicenna propose regarding the number of immaterial substances and celestial movements?

Avicenna proposes that the immediate end of celestial movements is not the first cause (God), but an intermediate intelligence. This implies that the number of immaterial substances could be greater than the number of celestial bodies, thus correcting the Aristotelian interpretation.

42-How is Aristotle's argument about the eternity of movement defended in light of Christian faith?

Although Aristotle bases his argument on the eternity of movement, Thomas Aquinas points out that its validity is not affected if one disregards this eternity, as the uniformity of celestial movements indicates that the First Unmoved Mover (God) has the capacity to move in a constant and uniform manner, which is compatible with Christian doctrine.

43-How does Aristotle's conception of immaterial substances differ from the Christian idea of angels?

Aristotle does not directly address the existence of angels as spiritual beings with will and intelligence. His immaterial substances, such as the First Unmoved Mover, are metaphysical principles that explain the order and movement of the cosmos, without being personal beings like the angels in Judeo-Christian tradition.

44-How do Plato and Aristotle agree regarding the mode of existence of immaterial substances?
Both philosophers agree that the first being is the ultimate cause of all immaterial substances. Plato affirms that the One is the cause of the unity and goodness of lower substances, similar to the sun that illuminates the air. Aristotle, on the other hand, holds that the First Unmoved Mover is the cause of the being and truth of all other things.

45-What is the main difference between Plato and Aristotle regarding the nature of immaterial substances?
Although both believe that immaterial substances are free from matter, Plato holds that these substances are composed of potentiality and actuality, whereas Aristotle asserts that the first being is pure actuality, without any mixture of potentiality.

46-How do Plato and Aristotle understand providence in relation to the first being?
Plato teaches that the One, being Goodness itself, cares for all inferior things through a providential hierarchy. Aristotle also accepts the existence of a separate and good being that provides order in the universe, but his approach is more teleological and less personal than Plato's.

47-What is the role of the First Unmoved Mover according to Aristotle in the hierarchy of the universe?
The First Unmoved Mover is regarded by Aristotle as the ultimate cause of movement and order in the universe. Although it does not directly intervene in the details of the world, it establishes a hierarchical order by being the ultimate end toward which all beings tend.

48-How do Plato and Aristotle's views on good and truth differ?

For Plato, the good and the truth are linked to the pure act of the Ideas or Forms, with the Supreme Good being the Form of the Good that guides everything toward perfection. Aristotle, in contrast, relates the good and truth to the natural purpose or end of beings, defining good as the realization of the natural end and truth as the correspondence with the essential nature.

49-What distinguishes participation in immaterial substances in Plato's conception from the Aristotelian one?

Plato understands participation as an ontological matter where material entities acquire their properties by participating in the Ideas. Aristotle sees participation in terms of efficient causality, where material entities receive their form and existence from efficient causes.

50-How does Aristotle apply the distinction between potentiality and actuality in his philosophy?

Aristotle uses the distinction between potentiality and actuality to explain change and development in a wide range of phenomena, from biology to metaphysics. Potentiality is the capacity of a being to change or receive a form, while actuality is the effective realization of that potential.

51-How does Plato's distinction between potentiality and actuality differ from Aristotle's?

In Plato, the distinction between potentiality and actuality is connected to the theory of Forms, where the sensible world is an imperfect copy of perfect Forms. The distinction is not as central as it is in Aristotle, who uses it to explain change and the realization of the end in a broader and more systematic sense.

52-How does Plato approach cosmic providence in comparison to Aristotle?

Plato relates cosmic providence to participation in the Supreme Good, which guides and organizes the universe through a hierarchy. Aristotle, on the other hand, presents a teleological view where providence is linked to the inherent purpose in the nature of beings.

53-What role do higher beings play in the cosmic hierarchy according to Plato?

In Plato's vision, higher beings, which are closer to the Supreme Good, possess greater perfection and have the ability to influence lower beings. These superior beings provide order and purpose to the lower levels of the hierarchy.

54-How does Aristotle view perfection and imperfection in the hierarchy of beings?

Aristotle believes that higher beings, due to their greater realization of their natural purpose, are more perfect and less flawed than lower beings. Lower beings, such as human souls, have more imperfections due to their connection with matter and its limitations.

55-How does Plato describe the hierarchy of separate substances in his philosophy?

Plato organizes separate substances into two main levels: intellects or intelligences, and the gods. The intellects, also known as *nous*, are capable of understanding the divine Ideas or Forms. The gods, as separate intelligible species, are eternal principles that enable the intellects to know. This hierarchy places the intellects below the gods and above the souls of celestial bodies.

56-What is the fundamental difference between Plato's and Aristotle's conceptions of separate substances?

Plato believes in a world of separate and perfect Ideas or Forms that exist independently of the physical world. Aristotle, in contrast, denies the existence of separate universals and holds that universal concepts are inseparably tied to individual things in the material world. Aristotle places the First Unmoved Mover at the peak of the order of substances, while Plato presents a dual order with intelligences and superior gods.

57-How does the number of inmaterial substances relate to celestial movements in Aristotle's philosophy?

Aristotle limits the number of immaterial substances to the number of celestial movements.

58-How does Plato approach the concept of intermediary souls in his cosmology?

Plato introduces the concept of intermediary souls, called daimons, who act as mediators between gods and humans. These souls are neither completely divine nor as material as human souls. Their function is to facilitate the communication and flow of divine influences between the higher and lower levels of reality.

59-What role does the First Unmoved Mover play in Aristotle's philosophy?

The First Unmoved Mover is the ultimate cause of all motion in the universe. It acts as the end or goal toward which all things strive, without itself moving or changing. Its knowledge is self-sufficient, based on its own essence, and it does not require the mediation of other beings to know or influence the cosmos.

60-How does Aristotle's approach to separate substances differ from Plato's?

While Plato grounds his separate substances in an intelligible world radically distinct from the sensible world, Aristotle presents a more integrated system in which separate substances, like the First Mover, explain the movement and order of the universe without the need for a separate world of Ideas.

61-Does Aristotle contemplate the idea of angels in his metaphysics?

Aristotle does not directly contemplate the existence of angels as understood in the Judeo-Christian tradition. His cosmology and metaphysics focus on concepts such as the first unmoved mover and separate substances, but he does not address the idea of angels.

62-What kind of being is the Aristotelian "first mover"?

The first moved mover is a being endowed with will and intelligence. The First Unmoved Mover is the immutable and pure principle of movement. It is not subject to motion or change and is the ultimate cause of motion in the universe. The first moved mover is the first heaven or the first celestial sphere that moves eternally due to the influence of the First Unmoved Mover.

63-Why is the Aristotelian "first mover" a body?

Saint Thomas says: *since nothing moves except a body, it must be concluded that the first mover is a body animated by an intellectual soul.*

64-How do human souls relate to intermediary souls and the souls of celestial bodies in Platonic cosmology?

In Platonic cosmology, human souls are further removed from the divine and intelligible world and are trapped in material bodies. Intermediary souls, or daimons, stand between celestial and human souls, acting as mediators. The souls of celestial bodies, being closer to the divine world, participate directly in the World of Ideas and have a higher role than human souls.

65-What does Avicebron assert about the composition of spiritual substances in his work *Fons Vitae*?

Avicebron asserts that all substances, including spiritual ones, are composed of matter and form. This principle, which Plato and Aristotle reserved only for material beings, is extended by Avicebron to the spiritual realm, positing a "spiritual matter" as the basis of immaterial substances.

66-What mistake does Avicebron make in confusing the concepts of potency and subject?

Avicebron confuses the concept of potency, which is the capacity to receive a form, with the concept of subject, which is the entity that possesses an actualized form. This error leads him to mistakenly believe that spiritual substances require material potency, like physical beings, when in reality they are pure forms.

67-What distinction does Avicebron make between the matter of celestial bodies and the matter of terrestrial bodies?
Avicebron distinguishes the matter of celestial bodies as a superior kind of matter, not subject to opposing qualities like cold or heat, making it more stable and perfect compared to the matter of terrestrial bodies, which is characterized by opposing qualities and is subject to change.

68-Why does Avicebron claim that no created substance can be completely simple?
Avicebron argues that no created substance can be completely simple because all creatures must be distinguished from the Creator, who is the only absolutely simple being. Due to this distinction, all created substances, both physical and spiritual, must be composed of matter and form.

69-Why does Avicebron consider spiritual substances to be finite?
Avicebron holds that the finitude of spiritual substances is due to their form, as it is form that defines and limits them. According to his argument, all created substances, including spiritual ones, must be composed of matter and form, and it is the form that establishes their finitude, in contrast to the infinite simplicity of the Creator.

70-What is the main criticism that Saint Thomas makes of Avicebron's position?
Saint Thomas criticizes Avicebron's idea that all substances, both corporeal and spiritual, are composed of matter and form. For Saint Thomas, spiritual substances, like angels and souls, cannot depend on matter, as their nature is predominantly actual and not potential, being closer to pure act.

71-What methodological error does Saint Thomas point out in Avicebron's explanation of higher realities?
Saint Thomas points out that Avicebron makes an error by attempting to explain higher realities using material principles. According to Saint

Thomas, the highest realities must be understood from their formal, not material, principles.

72- What does the primacy of act over matter mean for Saint Thomas?

The primacy of act over matter means that matter, on its own, is pure potentiality, that is, it has the potential to be but does not possess being in itself. Only when matter is united with form (act) can it acquire being. Therefore, act has ontological priority over matter.

73- Why does Saint Thomas consider the idea of a "universal matter" shared by both spiritual and corporeal substances inadmissible?

Saint Thomas considers the idea of "universal matter" inadmissible because it inappropriately reduces the spiritual to the material. Spiritual substances cannot be composed of matter since their nature does not depend on it, and equating them with material substances undermines their unique and superior character.

74- How does Avicebron's position affect the understanding of generation and corruption according to Saint Thomas?

For Saint Thomas, Avicebron's view destroys the understanding of generation and corruption by reducing all reality to a single matter. Matter, as pure potentiality, has no being of its own, and the true distinction between beings comes from forms, not merely from a common matter.

75- What is the problem with reducing the entire diversity of beings to a single matter, according to Saint Thomas?

Saint Thomas finds the reduction of all diversity of beings to a single matter problematic because it dilutes the essential differences between beings. Forms play a fundamental role in determining the essence of beings, and matter alone cannot explain the diversity of reality.

76- What critique does Saint Thomas make regarding the relationship between potentials and acts in Avicebron's theory?

Saint Thomas criticizes Avicebron for not correctly distinguishing the relationship between potentials and acts in spiritual and material substances. Avicebron suggests that all substances depend on a common matter.

77-How does Saint Thomas understand the distinction between spiritual and corporeal substances?
Saint Thomas understands that spiritual substances, like angels and souls, do not depend on matter for their existence, as they are self-subsistent forms. Corporeal substances, on the other hand, require the union of matter and form to exist, marking a radical distinction between the two.

78-Why does Saint Thomas consider that Avicebron's theory leads to logical contradictions?
Saint Thomas considers Avicebron's theory to lead to logical contradictions because, by postulating a "universal matter" for all things, it results in an infinite regression of material causes, which is inconsistent with fundamental metaphysical principles, such as the necessity of a first cause.

79-What aspect of the hierarchy of beings is compromised in Avicebron's theory according to Saint Thomas?
According to Saint Thomas, the hierarchy of beings is compromised in Avicebron's theory because he ignores that the highest in the order of beings, such as spiritual substances, cannot depend on something inferior like matter. In doing so, Avicebron confuses the essential distinction between the spiritual and the material.

80-What does Saint Thomas criticize in Chapter VII of the *Tractatus*?
Saint Thomas criticizes Avicebron's position on the common matter of bodily and spiritual substances, arguing that this is impossible due to the nature of matter and its relationship with forms.

81- Why does Saint Thomas maintain that matter cannot be the same for bodily and spiritual substances?

Saint Thomas maintains that matter cannot be the same for both types of substances because spiritual substances do not have physical dimensions, while bodily substances do. Matter must adapt to the nature of forms, and spiritual and bodily substances have different forms and characteristics.

82- How does bodily matter receive form according to Saint Thomas?

Bodily matter receives form in a "particularizing" manner, meaning that the form individualizes in matter according to its physical properties and limitations. This reception is limited and not total due to the constraints of bodily matter.

83- What distinguishes the reception of form in spiritual substances?

Spiritual substances receive form completely and without limitations, as they are not subject to physical or quantitative constraints. This allows them to understand and encompass the form in its entirety.

84- What does the complete capacity to receive form imply for spiritual matter?

The capacity of spiritual substances to receive form completely implies that their matter, if it existed, must be fundamentally distinct and higher than bodily matter, which is subject to physical limitations.

85- How does Saint Thomas relate potency and act in spiritual and bodily substances?

Saint Thomas relates potency and act by stating that spiritual substances are closer to act (complete realization) and further from potency (the capacity to become something). This means that spiritual substances are more perfect because their matter is closer to act.

86- What conclusion does Saint Thomas draw about spiritual matter when considering it as an entity in act?

Saint Thomas concludes that in spiritual substances, there is no real distinction between matter and form. This makes these substances simple and not composite, as matter and form are identified in a single reality.

87-What is the implication of the absence of a real distinction between matter and form in spiritual substances?
The absence of a real distinction between matter and form implies that spiritual substances are simple and complete in their existence, not composed of material parts like bodily substances.

88-What does Saint Thomas assert about the relationship between act and potency in the context of spiritual substances?
Saint Thomas asserts that act is superior to potency in the scale of being. Spiritual substances, being closer to act and more perfect, would have matter that is closer to act and less subject to potency.

89-What consequences does Avicebron's theory have according to Saint Thomas?
Saint Thomas considers that Avicebron's theory leads to an infinite regression in the search for material principles, denies the distinction between substance and accidents, and destroys the fundamental principles of metaphysics and natural philosophy.

90-What distinction does Saint Thomas make between celestial matter and elemental matter?
Saint Thomas distinguishes celestial matter, which is in potency for a perfect act and is stable, from elemental matter, which remains in potency for different forms and is subject to changes and transformations.

91-What position did some philosophers adopt regarding the existence of spiritual substances?
Some philosophers defended that spiritual substances, such as the soul or angels, are uncreated and have no cause. They believed that, not being subject to the material processes of change and movement, these

substances must exist independently and were not created by any entity, not even by God.

92-What is the main fallacy of the view that holds that spiritual substances cannot be created?
The main fallacy is the limited interpretation of creation, which assumes that all change or creation must follow the patterns of the physical world. This view ignores the possibility that immaterial substances can also be created by God, though in a manner different from material beings.

93-How does the position of those who believe in a hierarchical cause of spiritual substances differ from the Christian doctrine?
The hierarchical position holds that only the highest spiritual substances come directly from God, while the lower ones derive their existence from the higher ones. The Christian doctrine, in contrast, teaches that God is the direct cause of all spiritual substances, without intermediaries, regardless of their hierarchical level.

94-What philosophical error do proponents of the partial creation of spiritual substances commit?
They commit the error of limiting God's direct creative action by accepting that only the being of spiritual substances comes from God, while certain essential attributes or faculties of these substances would come from other superior creatures. This restricts God's direct influence in the creation of all facets of spiritual substances.

95-What advancement in philosophy allowed for a better understanding of the change and transformation of substances?
The advancement was the introduction of the distinction between matter and form. Philosophers began to understand that change does not only occur through the aggregation or disaggregation of material parts but also involves a transformation of the forms present in matter.

96-How does Christian doctrine explain the origin of being compared to pagan philosophers?

Christian doctrine explains that the origin of being does not depend on a mere physical or material change, but on the uncreated Being Himself, God. He shares His being with all creation. God is the cause of the being of all things, without the need for movement or mutation.

97-What analogy does Saint Thomas use to illustrate how God grants being to all things?
Saint Thomas uses the analogy of light. Just as a luminous body illuminates without altering its own nature, God illuminates the being of all things without subjecting Himself to mutation or movement. Similarly, intellectual truths are always true without the need for change.

98-What is the difference between the conception of spiritual substances by Plato and Aristotle and the Christian view of their creation?
Plato and Aristotle asserted that some substances have eternal existence and do not require temporal creation. In contrast, the Christian faith holds that all substances, even the eternal ones, were created *ex nihilo* by God, that is, from nothing.

99-What does the Angelic Doctor assert about the act of God's creation?
The Angelic Doctor asserts that God does not create out of internal necessity, but through a free act of His will and wisdom. God grants existence and duration to all things according to His eternal design, without being limited by time or acting in a forced or automatic manner.

100-How does Aristotelian-Thomistic philosophy relate to the explanation of the origin and nature of spiritual substances?
Aristotelian-Thomistic philosophy offers a complete and satisfactory explanation of the origin and nature of spiritual substances by holding that all things have their origin in God, who is the First cause. This view includes the idea that creation does not require material change, and spiritual substances, although eternal, can be created by God without the need for mediators.

101-What criticism does Saint Thomas make of Avicenna's hierarchical view of creation?

Saint Thomas criticizes Avicenna's hierarchical view for implying that the First Principle, God, does not act directly in creation. Instead, Avicenna suggests that creation follows a chain of secondary causes. Saint Thomas argues that the order and perfection of the universe must be the result of God's direct intention, not of a chain of intermediaries.

102-What is the difference between production by movement and creation without movement according to Saint Thomas?

Saint Thomas differentiates between production by movement, which involves a process of change and depends on secondary causes, and creation without movement, which is a direct and absolute act of God. Creation without movement refers to the production of immaterial substances and celestial bodies, and does not involve change or movement.

103-How does Saint Thomas define universal cause in contrast with particular cause?

The universal cause is what produces nature or form in general, applicable to all subjects of a species, while the particular cause produces a form in a specific subject. God is the universal cause that produces the nature of all things, whereas a particular cause, like a horse, transmits the equine nature to a specific individual.

104-What does the infinite virtue of God imply according to Saint Thomas?

The infinite virtue of God implies that He has the absolute capacity to create without depending on anything preexisting. Unlike created beings, whose virtue may be finite or limited, God can cause anything from nothing *(creatio ex nihilo)* due to His infinite virtue.

105-Why, according to Saint Thomas, must immaterial substances be created directly by God?

Saint Thomas holds that immaterial substances must be created directly by God because only God's infinite virtue can produce such substances without the intervention of secondary causes or processes of movement.

106-What arguments does Saint Thomas present against the idea of a first intelligence created?

Saint Thomas argues that the notion of a first intelligence created limits divine causality and does not adequately explain the origin of cosmic order. He asserts that the creation and order of the universe must be the direct result of divine intention, not of a chain of secondary causes.

107-How does Saint Thomas distinguish between creation and generation?

Saint Thomas distinguishes between creation and generation by noting that creation is the act of bringing something into existence that did not previously exist, while generation involves a change of state in something already existing. Creation is exclusive to God and does not involve movement, unlike generation.

108-What role does God play according to Aristotelian-Thomistic philosophy in the context of creation?

In Aristotelian-Thomistic philosophy, God is the First and only cause of all substances. Creation is a direct and instantaneous act of God, who acts as the universal cause of all things, reflecting His intention and perfection in the order of the universe.

109-What role do secondary causes play in Saint Thomas's view of creation?

According to Saint Thomas, secondary causes can only transmit forms and cannot produce being itself. The creation of being itself is an exclusive gift from God, and secondary causes operate within the framework established by God's direct action.

110-How does creation *ex nihilo* relate to Saint Thomas's view on the origin of the universe?

Creation *ex nihilo*, or creation from nothing, is central to Saint Thomas's view on the origin of the universe. According to him, God creates all substances, both material and spiritual, from nothing, underscoring that the universe directly reflects divine will and perfection without relying on a gradual process or secondary causes.

111-What is the Neoplatonic view on causality according to Chapter XI?

Neoplatonists believe that God is the immediate cause of being for all immaterial substances. This means that God directly creates all spiritual entities, such as the soul and intellect, without the need for intermediaries. Creation is understood as a pure and direct act of divine will, without involving mutation or physical movement.

112-How do Neoplatonists organize abstract principles in their doctrine?

Neoplatonists organize abstract principles in a hierarchical order. The most general principles, such as the "One" and "being," are the first and most universal, followed by the principle of "life," and finally, the principle of "intellect." This hierarchy indicates that each principle contributes different aspects to spiritual substances.

113-What criticism does Saint Thomas make of the Neoplatonic doctrine regarding separated principles?

Saint Thomas criticizes the Neoplatonic doctrine for suggesting that the essential qualities of a substance (such as being, living, and understanding) come from various distinct principles. According to him, all these qualities must proceed from a single principle, which is God.

114-How does Saint Thomas refute the idea of the separation of principles in immaterial substances?

Saint Thomas refutes the Neoplatonic idea that being, life, and understanding can be separate principles added to the essence of a substance as if they were accidents. He argues that in immaterial substances, these qualities are not independent of each other, but all originate from the same principle: God.

115-What partial acceptance does Saint Thomas make of the Neoplatonic opinion?
Saint Thomas partially accepts some aspects of Neoplatonic theory, such as the possibility that certain additional aspects, like "intelligible species," may derive from a hierarchy of superior substances according to the order established by God. However, he rejects the idea that the essential qualities themselves come from separated principles.

116-What is Saint Thomas's main criticism of Origen's position on the creation of spiritual substances?
Saint Thomas criticizes Origen's position that all spiritual substances were created equal and that diversity arose later due to the free choice of these substances. Thomas considers this idea inconsistent because it does not reflect divine perfection. For him, the perfection of the universe requires diversity and inequality from the beginning, as these characteristics contribute to a divine order and hierarchy that cannot arise randomly.

117-Why does Saint Thomas argue that initial inequality is necessary for the perfection of the universe?
Saint Thomas argues that the perfection of the universe is not based on uniform equality among spiritual substances but on their diversity and order. According to him, a universe composed solely of identical entities would be imperfect and lack the complexity and harmony needed to achieve perfection. Diversity and inequality, from the beginning, are essential for a cosmos that adequately reflects divine perfection.

118-How does Saint Thomas view the relationship between divine justice and inequality in creation?
Saint Thomas holds that the creation of unequal spiritual substances is not unjust. True divine justice manifests in an ordered and diverse universe, where each thing has a specific role and contributes to the divine plan.

Diversity and inequality in creation are necessary for the well-being and perfection of the cosmos, as they allow each element to fulfill a function within the divine order.

119-What criticism does Saint Thomas make of the idea that inequality and the creation of bodies result from the free will of spiritual substances?

Saint Thomas criticizes Origen's idea that inequality and the creation of bodies result from the misuse of the free will of spiritual substances. Thomas argues that this view attributes the order of the universe to chance or disordered decisions, rather than recognizing that order and hierarchy have been present from creation, reflecting divine perfection. Creation cannot be the result of disobedience, but of a divine plan that already incorporates an intrinsic order.

120-What is Saint Thomas's perspective on the necessity of multiplicity in spiritual substances?

Saint Thomas argues that, unlike physical bodies that may need multiplicity to fulfill specific functions, spiritual substances, being more perfect, do not require multiplicity to achieve their purpose. The existence of multiple instances of spiritual substances at the same level of perfection would be unnecessary and would indicate a lack of intrinsic perfection. For Thomas, a perfect spiritual substance does not need multiplicity to fulfill its function in the universe.

121-How does Saint Thomas address the concept of "free will" in relation to the creation and inequality of spiritual substances?

Saint Thomas holds that the free will of spiritual substances is not the factor that originates inequality in creation. In his view, free will cannot explain creation itself nor the hierarchy established from the beginning. For him, inequality and order in the universe are part of the initial divine design, not the result of decisions made freely by spiritual substances. Free will may influence individual behavior, but not the fundamental structure of the cosmos.

122-What role does divine perfection play in explaining inequality according to Thomas Aquinas?
Thomas Aquinas argues that divine perfection is manifested not through uniform equality but through diversity and hierarchy. Divine perfection requires a universe where each being has a specific place and function, contributing to a general order that reflects God's perfection. Inequality in creation is, therefore, an expression of this perfection, as it allows for a more complete and harmonious organization of the cosmos.

123-How is the concept of "hierarchy" related to Thomas Aquinas's criticism of Origen's view?
For Thomas Aquinas, hierarchy is essential for the perfection of the universe. Origen's idea, which suggests that inequality arises only after creation, fails to recognize that hierarchy is an integral part of the divine plan from the beginning. Aquinas maintains that hierarchy is not an accident but a fundamental component of the order created by God, reflecting His wisdom and perfection.

124-What implications does Thomas Aquinas's view on the nature of spiritual substances have for understanding their role in the universe?
Aquinas's view implies that spiritual substances have defined roles within the universe, based on their position in a hierarchy established by God. Diversity and inequality are not seen as defects but as necessary aspects that allow each substance to fulfill a particular function within the cosmos. This contrasts with Origen's idea that inequality is the result of free and disorderly choices.

125-How does Saint Thomas defend the creation of bodies and the diversity in the universe as part of the divine plan, in contrast to the idea that they are a "prison" for souls?
Aquinas rejects the idea that the creation of bodies is a form of imprisonment for souls that have deviated from their original state. Instead of viewing the creation of bodies as a restriction, Aquinas considers that bodies and the diversity in the universe are part of the divine plan to

achieve greater perfection. The existence of bodies is not a punishment but a part of God's ordered and perfect design that allows for a complete manifestation of both spiritual and material reality.

126-What does Thomas Aquinas consider in Chapter XIII regarding the knowledge of God and spiritual substances?

Aquinas addresses errors concerning the knowledge and providence of God, as well as the knowledge of spiritual substances, which arise from incorrectly comparing these substances to human capacities and operations.

127-Why did some thinkers believe that God and angels only knew universals?

Some thinkers believed that since God and angels do not possess bodies or senses, they could not know singulars (particular individuals), as in the case of humans, knowledge of singulars is achieved through the senses.

128-How does Thomas Aquinas refute the idea that God and angels cannot know singulars?

Aquinas argues that God and angels are not limited like humans and possess a much more perfect knowledge. They can know both universals and singulars directly, without relying on the senses.

129-Why is it incorrect to limit God's knowledge to universals according to Thomas Aquinas?

It is incorrect because God has absolute and perfect knowledge that encompasses everything that exists, including singulars. Limiting His knowledge to universals underestimates His perfection and superior nature.

130-What error did some thinkers commit regarding divine knowledge?

They held that if God knew something external to Himself, it would perfect Him, which would be contradictory since God is already perfect. Therefore, they argued that God can only know His own essence.

131-How does St. Thomas respond to the error regarding divine knowledge?

St. Thomas maintains that God, being omniscient, knows everything that exists without this implying that He becomes perfected. Furthermore, as the cause of all, God knows creatures without depending on them to acquire knowledge.

132-What error arises regarding the necessity of future events under divine providence?

Some thought that if everything is under divine providence, future events must happen necessarily, eliminating human freedom and contingency.

133-How does St. Thomas reconcile divine providence with human contingency and freedom?

St. Thomas argues that God knows future events, including contingent ones, without imposing necessity upon them. Divine providence encompasses everything but respects the freedom and contingency of human actions.

134-How does St. Thomas explain the relationship between evil and divine providence?

St. Thomas asserts that evil is permitted within the order of the universe, and God allows it not because He causes it, but because a greater good can arise from it. Evil does not contradict divine providence.

135-What view does St. Thomas reject regarding the providence?

St. Thomas rejects the dualistic idea that divine providence does not encompass the lower world or human events. He argues that divine providence covers the entire universe, both spiritual and material.

136-What fundamental aspects of divine nature does St. Thomas explore in Chapters XIV and XV?

St. Thomas explores the knowledge and providence of God.

137-How does St. Thomas address the issue of divine knowledge in his text?

St. Thomas divides his argument into two main parts: he examines the nature of divine knowledge and divine providence.

138-What does St. Thomas assert about divine knowledge and its relationship to the substance of God?

St. Thomas asserts that in God, substance and understanding are identical, implying that His knowledge is absolute and universal.

139-How does St. Thomas use the idea of separated forms to explain divine understanding?

St. Thomas compares divine understanding to separated forms, explaining that just as a separated form is complete in itself, God's understanding is also perfect and universal.

140-What difference does St. Thomas establish between the participated knowledge of finite beings and the essential knowledge of God?

Finite beings have knowledge through participation, conditioned by their limited nature, while divine knowledge is essential, absolute, and not conditioned by external factors.

141-How does St. Thomas describe the knowledge God has of His own substance?

St. Thomas asserts that God perfectly knows His own substance, reflecting the perfection of His being and encompassing all possibilities of existence.

142-What is the notion of God as the first cause according to the philosophy of St. Thomas Aquinas?

God is considered the First Cause, the fundamental principle of everything that exists, containing within Himself all principles and realities of effects.

143-How does St. Thomas criticize the Aristotelian interpretation of divine understanding?

St. Thomas criticizes the idea that divine understanding depends on external intelligible forms, arguing that divine knowledge must be absolute and independent.

144-What role does divine providence play in the order of the universe according to St. Thomas?

Divine providence organizes and directs the universe toward its ends, ensuring that everything in the cosmos unfolds according to the divine plan and purpose.

145-What is the First Unmoved Mover and how is it related to divine providence according to St. Thomas?

The First Unmoved Mover is God, who initiates and sustains movement in the universe without being moved by anything external. His providence guides all actions and events toward their ends.

146-What does Thomas assure about the perfection of order in the universe?

Thomas assures that the order of the universe is not the result of chance, but of an essential principle that reflects divine will and plan.

147-What are the two main aspects of the perfection of divine providence according to Thomas?

The two aspects are perfection in arrangement, which refers to the planning and ordering of the universe, and perfection in execution, which ensures that plans are implemented effectively and harmoniously.

148-How does Thomas explain the role of angels in the execution of divine providence?

Thomas explains that the execution of divine providence does not occur directly by God in all aspects of the universe; instead, God uses angels to carry out His plans and purposes.

149-Why is the idea of separated forms useful for understanding divine knowledge?
The idea of separated forms is useful because it shows how divine understanding, like a pure form, must completely and without limitations encompass all cognizable reality.

150-What does the absolute simplicity of God imply for His ability to know?
The absolute simplicity of God implies that His being and His capacity to understand are inseparable and coincide completely, reflecting an absolute and perfect knowledge of all that is knowable.

151-How does Thomas address the question of divine knowledge in relation to the criticisms and doubts it has raised?
Thomas defends that God possesses absolute and universal knowledge that encompasses all that is knowable, arguing that there is no limitation in divine knowledge. He explains that, being the Supreme being, God has a complete and perfect understanding of all things.

152-What does it mean that the divine substance and God's understanding are identical according to Thomas?
Thomas holds that, since God is absolutely simple, His essence and His capacity to understand are not distinct but one and the same. This identity between substance and understanding implies that divine knowledge is total and perfect, encompassing all that is knowable.

153-How does Thomas use the idea of separated forms to explain divine understanding?
Thomas resorts to separated forms as an analogy for divine understanding, suggesting that, just as a separated form like "whiteness" contains within itself the fullness of its essence, divine understanding must completely encompass all cognizable reality.

154-What distinguishes participated knowledge of finite beings from the essential knowledge of God?

The participated knowledge of finite beings is derived and conditioned by their limited nature, while the essential knowledge of God is absolute, inherent to His essence, and does not depend on external factors. God possesses complete knowledge independent of any conditioning.

155-What does it imply for St. Thomas that God knows His own substance perfectly?

It means that God has absolute and total knowledge of His own nature and power. His knowledge of His substance includes all possibilities and capacities derived from it, encompassing all that is possible within His infinite power.

156-How does Thomas explain the relationship between the First Cause and divine knowledge?

God, as the First Cause, contains within Himself all the principles and realities found in effects. This implies that His knowledge includes all the characteristics and possibilities of everything that exists, making His knowledge absolute and complete.

157-What is Thomas's criticism of the Aristotelian interpretation of divine understanding?

Thomas criticizes the Aristotelian idea that divine understanding depends on external intelligible forms, arguing that this would imply that divine knowledge is not absolute or independent. According to Thomas, divine understanding must be autonomous and reflect the perfection of divine substance.

158-How does Thomas extend the concept of divine providence to all that exists?

Thomas holds that, just as God has absolute knowledge of all things, His providence also encompasses the entire universe. This means that divine providence directs and governs all aspects of creation toward their respective ends, ensuring that the universe functions according to the divine plan.

159-What role does the concept of the First Unmoved Mover play in Thomas's view of God?

God is the First Unmoved Mover who initiates and sustains movement and action in the universe without being moved by anything external to Himself. This concept underscores that God is the primary and supreme Cause of everything that occurs in the cosmos, influencing all levels of reality.

160-How does the perfection of divine providence manifest according to Thomas?

The perfection of divine providence manifests in the perfect arrangement and execution of God's plan. The arrangement involves complete and orderly planning, while the execution refers to the efficient and universal implementation of the divine plan.

161-What role do angels play in the execution of divine providence according to Thomas?

Thomas explains that angels act as executors of divine providence, carrying out actions and events in the world according to God's plan. This function highlights the cooperation between the divine and the created in fulfilling divine order and purpose.

162-How does Thomas respond to the objection regarding the ability of divine understanding to know singular individuals?

Thomas explains that higher beings, such as angels and God, possess a more universal cognitive virtue, allowing them to know both the universal and the singular, unlike limited human understanding.

163-Why does Thomas maintain that higher beings can know singular individuals?

He argues that the knowledge of higher beings reflects the order and structure of beings in the world, with a cognitive virtue that is broader and more universal.

164-How does human understanding become perfected according to Saint Thomas?
Human understanding is perfected by receiving abstract intelligible forms from sensible objects, enhancing its knowledge through the perception of forms.

165-Why does divine understanding not need to be perfected?
Divine understanding is perfect in itself because its knowledge is part of its own essence; it does not need to receive external forms to be perfected.

166-What distinction does Saint Thomas make between the knowledge of humans and that of angels?
Angels derive their perfection from divine knowledge, while human understanding is perfected through the reception of intelligible forms from sensible objects.

167-How does Saint Thomas address the objection concerning random events and divine providence?
Saint Thomas explains that what seems random from the human perspective is actually ordered within the perfect plan of divine providence, which encompasses all events in the universe.

168-What example does Saint Thomas use to illustrate the difference between lower and higher agents in relation to random events?
He uses the example of a person sent to a place where they know there will be thieves; the assault seems casual to the messenger, but it is not for the one who planned it.

169-What distinction does Saint Thomas make between natural causes and rational causes?
Natural causes produce effects according to their intrinsic nature, while rational causes, like humans, produce effects based on plans and concepts.

170-How does Saint Thomas reconcile the existence of evil with the supreme goodness of God?

He explains that evils arise from the limitations of secondary causes, not from God's direct action. God permits evil in His divine plan to achieve a greater good.

171-Why does God allow the existence of evil according to Saint Thomas?

God allows evil because in His divine plan, these evils can contribute to a greater good or to the overall balance of the universe.

172-What fundamental error do the Manicheans make regarding cosmic principles?

The Manicheans make the error of believing in two opposing and coeternal cosmic principles that explain the origin and structure of the universe: one for good and the other for evil. This contradicts the view of a single creator principle, such as God in Christianity.

173-How do the Manicheans describe the principles of good and evil?

The Manicheans describe the principle of good as an infinite corporeal light with intellect, and the principle of evil as an infinite corporeal darkness. They believe that both principles have a physical and material nature.

174-What is the Manichean view on the nature of evil?

The Manicheans maintain that evil has a physical and substantial existence, just like good. They believe that evil is a corporeal entity represented by darkness.

175-How does Saint Thomas refute the idea that evil can be an active principle?

Saint Thomas refutes the idea that evil can be an active principle by arguing that evil has no active existence in itself. Evil is a privation of good or perfection and, therefore, cannot produce effects by itself.

176-What example does Saint Thomas use to illustrate the inability of evil to act?

Saint Thomas uses the example of the relationship between body and soul. The soul is the perfection and act of the body. Death, which is considered an evil, is not an active force but rather the privation of the soul, illustrating that evil does not produce anything by itself.

177-What argument does Saint Thomas present against the idea that the intellect can be corporeal?

Saint Thomas argues that the intellect cannot be corporeal because, if it were, it would be limited by the physical characteristics of that matter and could not know all things. The intellect, being immaterial, transcends the limitations of space and time.

178-Why is it incorrect to attribute a corporeal nature to the principles of good and evil according to Saint Thomas?

According to Saint Thomas, attributing a corporeal nature to the principles of good and evil is incorrect because good, being an intellectual principle, cannot be material. A body has physical limitations that the intellect, in its immaterial nature, does not have.

179-How does Saint Thomas understand the role of good in the governance or direction of the universe?

Saint Thomas holds that good is the objective towards which all things tend and is the end of all governance or direction. There cannot be a "kingdom of evil" as an end in itself, since evil is a privation of good and cannot be a final objective.

180-What error do the Manicheans make when extrapolating observations from the material world to universal principles?

The Manicheans err in extrapolating the observation that contrary effects come from contrary causes in the material world to the realm of first principles. They mistakenly believed that good and evil, as contraries, must be active principles of all things.

181-What is Saint Thomas' conclusion regarding the existence of the principles of good and evil?

Saint Thomas concludes that contraries are not first or creator principles but must have a common superior cause. This common cause is God, who is the only creator principle, and the contraries derive from this one superior cause.

182-What does the Christian doctrine hold about the origin of spiritual substances?

The Christian doctrine holds that spiritual substances, such as angels, were created directly by God. This belief aligns with the idea that both the physical and spiritual worlds are products of God's creative action.

183-How do the Psalms reinforce the idea that spiritual substances were created by God?

Psalm 148, 2 mentions angels and other heavenly powers, indicating that they should praise God. Psalm 148, 5 underscores that all things, both physical and spiritual, were created by the Word of God, reinforcing the notion that everything is the work of divine creation.

184-What does Pseudo-Dionysius the Areopagite argue about spiritual substances and the goodness of God?

Pseudo-Dionysius the Areopagite argues that spiritual substances, including angels, are a direct product of God's goodness. According to him, God, by His universal goodness, gives existence to the essences of all things, including spiritual substances.

185-What is the position of Pseudo-Dionysius the Areopagite on the independent existence of spiritual substances?

Pseudo-Dionysius the Areopagite maintains that spiritual substances do not have an independent existence or one generated by other beings. Instead, all of them come directly from divine goodness, which is the unique source of their being and perfection.

186-What does Pseudo-Dionysius the Areopagite assert about the relationship of spiritual substances with divinity?

Pseudo-Dionysius the Areopagite asserts that spiritual substances are not created by a multiplicity of distinct divine causes or principles. All their qualities, such as goodness, being, and life, originate from the unique and absolute goodness of God.

187-How does Pseudo-Dionysius the Areopagite refute the Platonic view of spiritual substances?

Pseudo-Dionysius the Areopagite refutes the Platonic view by asserting that there are not multiple gods or divine principles responsible for different aspects of spiritual substances. Instead, he holds that there is only one God who is the source of all that is good, true, and life-giving.

188-What is the Christian teaching regarding the existence of spiritual substances according to the text?

Christian teaching maintains that spiritual substances have not existed eternally, but were created by God at a specific moment in time. This view contrasts with Platonic philosophy, which proposes an eternal existence for spiritual substances.

189-What do St. Augustine, Gregory of Nazianzus, and St. Jerome think about the creation of angels?

St. Augustine believes that angels were created simultaneously with corporeal creatures. Gregory of Nazianzus and St. Jerome believe that angels were created before corporeal creatures. Additionally, some suggest that angels were created in light, symbolizing purity and clarity.

190-How is the concept of place related to the creation of angels according to the text?

If angels were created before corporeal creatures, they cannot be assigned a physical place because place is associated with the corporeal. If they were created at the same time as corporeal creatures, some theologians suggest that they may have been created in a "luminous heaven," a spiritual state reflecting their purity and relationship with divine light.

191-What is understood by the presence of angels in heaven according to the text?

The presence of angels in heaven should not be interpreted as a physical location in a material sense, but as a spiritual presence. Angels are present in heaven in a spiritual way, through a virtual or relational connection, rather than through a corporeal location.

192-How does Pseudo-Dionysius the Areopagite describe the nature of angels?

According to Pseudo-Dionysius the Areopagite, angels are incorporeal and immaterial. Although material figures are used to illustrate the hierarchy and nature of angels, these images are merely symbolic representations that help to understand the spiritual reality of angels.

193-What is St. Augustine's position regarding the corporality of angels?

St. Augustine considers angels to be spiritual and intellectual entities, without material corporality. However, he acknowledges that, compared to the spiritual purity of God, the incorporeality of angels may still seem material.

194-According to St. Basil, where are the angels located and how is their state described?

St. Basil describes angels as being in light and spiritual joy. This indicates that their location should not be understood physically, but as an elevated spiritual state in which they experience divine fullness and joy.

195-What does it mean that angels are in "intelligible places" according to St. Gregory of Nyssa?

St. Gregory of Nyssa states that angels are in intelligible places, meaning that their existence relates to places corresponding to their spiritual and intellectual nature, rather than physical locations. These "places" are understood in terms of their relationship with intelligence and spiritual reality.

196-How should the bodily representations of angels in Scripture be interpreted?

The bodily representations of angels in Scripture should be understood allegorically. Physical descriptions do not indicate that angels have material bodies, but rather serve as metaphors to aid in the understanding of spiritual realities.

197-What does the use of terms like "anger" or "concupiscence" imply when referring to angels?

Terms like "anger" or "concupiscence" applied to angels should be understood symbolically. They do not indicate that angels experience human emotions, but reflect the manner in which angels act in accordance with divine law and their spiritual functions.

198-How should the movement of angels be understood?

The movement of angels should be understood spiritually, not in corporeal terms. Angels do not move physically, but their "movement" symbolizes their spiritual activity and influence in various areas of the spiritual realm.

199-What is the text's approach regarding the presence of angels in physical places?

Angels are not present in physical places materially. Their presence in a place refers to a spiritual or relational connection with that place, not to a tangible physical location.

200-What is the main focus of Chapter XX?

Chapter XX focuses on the distinction between angelic spirits, particularly the difference between good and evil spirits, and examines various opinions about the nature of demons and the reasons why some angels became evil.

201-How does St. Thomas describe good spirits and what function do they serve?

St. Thomas describes good spirits as "ministering spirits" who serve God and assist humans destined for salvation. He cites Hebrews 1, 14 to support this assertion, where it is mentioned that these spirits are sent to serve the faithful and just.

202-What characteristics does St. Thomas attribute to evil spirits according to Matthew 12, 43?

According to Matthew 12, 43, evil spirits are described as "unclean spirits" that seek rest but group together with worse ones. Jesus mentions that an unclean spirit leaving a person seeks a place of rest and, upon finding none, returns to discover the house (the person) in worse condition, bringing worse spirits with it.

203-What is St. Thomas's opinion on the existence of evil?

St. Thomas maintains that evil does not have its own existence; it is a deprivation of good. Evil occurs when something good is absent or altered, and therefore, demons cannot be considered evil by nature, as evil is a corruption of good existence.

204-How does St. Thomas address the concept of corruption by demons?

St. Thomas argues that demons do not have the capacity to corrupt everything they touch or even to corrupt themselves entirely. This reinforces his viewpoint that evil is not an inherent characteristic of demons.

205-What does St. Thomas say about the creation of demons by a good God?

According to St. Thomas, God is good and creates only good things. If demons were inherently evil, it would imply that a good God created something evil, which would be contradictory. The existence of demons cannot be attributed to the creation of something inherently evil.

206-What distinction does St. Thomas make between the natural inclinations of corporeal beings and incorporeal beings?

St. Thomas argues that natural inclinations toward evil are characteristics of corporeal beings, which are subject to passions and material desires. In contrast, incorporeal beings, such as demons, do not have these natural inclinations toward evil derived from a physical existence and, therefore, should not have a natural inclination toward evil.

207-What is St. Thomas's position on the theory of demons as corporeal beings?

St. Thomas rejects the theory that demons are corporeal beings. Instead, he holds that demons are spiritual beings, originally angels created by God, who became demons through their free will by rebelling against God.

208-How does St. Thomas use Scripture to support his argument about the fall of angels?

St. Thomas cites Isaiah 14, 12-15 and Ezekiel 28, 12-17 to demonstrate that fallen angels became evil through their own transgression. Isaiah describes the fall of a proud being aspiring to rise above God, while Ezekiel speaks of a being perfect in beauty and wisdom who was corrupted by pride. These passages support the idea that evil in demons arises from their free decisions and rebellion against God.

209-What is St. Thomas's opinion on the Platonic idea of the transformation of human souls into demons?

St. Thomas rejects the Platonic idea that human souls can transform into demons. According to him, demons are fallen angels who became evil through their free choice and rebellion against God, and human souls have a distinct nature and cannot become demons.

210-How does St. Thomas explain the cause of the rebellion of some angels?

St. Thomas explains that the rebellion of some angels was caused by their free will and pride. The fallen angels, led by Lucifer, rebelled against God due to their desire to be equal to God and their refusal to accept the divine hierarchy established by God. This rebellion was not prompted by a natural inclination toward evil, but by a free and conscious choice.

211-What role do demons play in the history of salvation according to St. Thomas?

According to St. Thomas, demons play the role of tempters and opposers in the history of salvation. Their function is to tempt human beings and try to divert them from the path of salvation. Despite their opposition, their influence is limited by divine will, and ultimately, God's purpose is for evil and temptations to be used for good and to strengthen the faith of the faithful.

212-How does Saint Thomas address the relationship between the free will of the angels and their fall?

Saint Thomas argues that the free will of the angels is essential to understanding their fall. The angels had the freedom to choose between obeying God or rebelling against Him. The fall of the angels was not the result of an inevitable necessity, but of a free choice in which they decided to rebel against divine authority.

213-What differences does St. Thomas establish between the nature of angels and that of demons?

St. Thomas distinguishes between angels and demons in terms of their essential nature. Both are spiritual beings created by God, but angels remain faithful and continue to serve God, while demons, who were fallen angels, have chosen to rebel against God. The difference lies in the free choice of demons, which led them to corruption and rejection of good.

214-How does the concept of "privation of good" relate to St. Thomas's view of evil?

The concept of "privation of good" in St. Thomas's view means that evil does not have its own existence, but is the absence or corruption of a good that should be present. According to St. Thomas, demons are not evil by nature, but have become evil through the privation of good, meaning they have chosen to reject good and, therefore, have become corrupt.

215-How does St. Thomas's view of demons compare with that of other theologians of his time?

St. Thomas's view of demons is distinguished from that of other theologians of his time in its emphasis on free will and the privation of good. While some contemporary theologians might focus more on the mythical or anthropomorphic aspects of demons, St. Thomas bases his view on a more philosophical and theological perspective, centered on the morality and spiritual nature of angels and demons.

216-What implications does St. Thomas's conception of demons have for the understanding of divine providence?

St. Thomas's conception of demons implies that divine providence is not disturbed by the existence of evil. Although demons act against God's plan, their power is limited by divine will, and everything that occurs, even evil and temptations, serves the ultimate fulfillment of divine purpose. Divine providence remains whole and sovereign despite the presence of evil.

217-How do St. Thomas's ideas about demons influence moral and spiritual theology?

St. Thomas's ideas about demons influence moral and spiritual theology by providing an understanding of temptations and evil that emphasizes the importance of free will and moral responsibility. His approach teaches the faithful that evil is not an intrinsic force in creation, but a corruption of good, and helps believers understand how to fight against temptations and maintain faith amid spiritual adversity.

ENDNOTES

[1] Cf. LESCOE FRANCIS J. *Treatise on Separate Substances: A Latin-English Edition of a Newly-Established Text Based on 12 Mediaeval Manuscripts with Introduction and Notes.* West Hartford, Connecticut: Saint Joseph College, 1963. https://catholiclibrary.org/library/view?docId=/Medieval-EN/XCT.028.html;chunk.id=00000003.

[2] DE AQUINO TOMAS SANTO. *Opúsculos filosóficos genuinos.* Según la edición crítica del P. Mandonnet OP. Introducción, notas explicativas y versión castellana por el Pbro. Antonio Tomás y Ballús. Editorial Poblet. Buenos Aires. 1947. Page 174.

[3] **Joseph Eschmann (1890-1968)** was a 20th-century Jesuit philosopher and theologian, known for his studies on Saint Thomas and scholastic philosophy. Eschmann stood out for his involvement in debates on natural law and his defense of Thomism in a contemporary context. He collaborated with influential figures such as Jacques Maritain, and his works addressed topics such as ethics, metaphysics, and philosophy of law.

[4] **Karl Henle (1890-1979)** was a German philosopher and theologian known for his work in scholastic philosophy and his focus on the thought of Saint Thomas Aquinas. As a member of the Society of Jesus, Henle significantly contributed to the interpretation and dissemination of Thomism in the 20th century, with a focus on metaphysics and ethics.

[5] **Étienne Mandonnet (1858-1936)**: A French Dominican historian and theologian specializing in the Middle Ages. He was a prominent scholar of the works of Saint Thomas Aquinas and scholastic philosophy. He is also known for his research on the mendicant orders and their influence on medieval education.

Louis Glorieux (1908-1991): A Belgian ecclesiastical historian and medievalist, expert in the works of Saint Thomas Aquinas and medieval scholasticism. He published important critical editions of medieval texts, significantly contributing to the study of Thomistic thought and its reception in the Catholic tradition.

Louis Walz (1890-1967): A German scholastic theologian and member of the Society of Jesus, specializing in the philosophy of Saint Thomas Aquinas. He was a strong advocate for Thomism during the 20th century and worked on the reinterpretation of his thought in contemporary contexts.

Hugh Callus (1900-1980): A Maltese Dominican theologian, specializing in dogmatic and scholastic theology. His work included detailed studies on the theology of creation and the relationship between divine grace and human freedom, primarily from a Thomistic perspective.

Martin Grabmann (1875-1949): A German theologian, historian of philosophy, and Catholic priest recognized for his research on medieval scholasticism. He is famous for his studies on Saint Thomas and his monumental work on the history of medieval theology, considered pioneering in its field.

[6]**Jean Vansteenkiste (1909-2003)**: A Belgian philosopher and theologian, member of the Society of Jesus, known for his research in scholastic philosophy. He was an advocate of neotomism and contributed to the dissemination of Saint Thomas's thought in the 20th century, with an emphasis on topics of ethics and metaphysics.

[7]DE AQUINO TOMAS SANTO. *Opúsculos filosóficos genuinos*. Según la edición crítica del P. Mandonnet OP. Introducción, notas explicativas y versión castellana por el Pbro. Antonio Tomás y Ballús. Editorial Poblet. Buenos Aires. 1947. Page 175.

[8]CALVET ANTOINE. *Thomas D'AQUIN, Les substances séparées*. Traduction, introduction et notes de Nicolas BLANC. *Revue de l'histoire des religions*, 3. 2019. Pages 619-622.

[9]CALVET ANTOINE. *Thomas D'AQUIN, Les substances séparées*. Traduction, introduction et notes de Nicolas BLANC. *Revue de l'histoire des religions*, 3. 2019. Pages 619-622.

[10]"Elements" refers to the basic components of matter according to the classical conception, for example: water, earth, air, fire. Saint Thomas distinguishes between two types of matter: celestial matter and elemental matter.

Celestial matter: The matter of celestial bodies is considered to be in potentiality for a perfect act. This means that this matter has the capacity to receive a form that completes all its potentiality, meaning there is no unfulfilled capacity left. The form it receives fully actualizes the potentiality of the matter. Therefore, celestial matter is not subject to change or transformation, as its form has already realized all its potential capacities. As a result of this perfection, celestial matter is stable and does not undergo change, having reached a state in which nothing remains to be perfected.

Elemental matter: In contrast, the matter of the elements remains in potentiality for different forms. This means that this matter has not attained a form that completes all its potentiality. The form it receives does not exhaust the matter's capacity to receive other forms. Therefore, elemental matter remains susceptible to change and transformation, as it retains remaining potentiality. Due to this incompleteness of its potentiality, elemental matter is subject to change and the possibility of receiving new forms.

This distinction allows for an understanding of how matter can vary in its characteristics and capacities depending on its degree of perfection. Celestial matter represents an idealized and perfected type of matter, while elemental matter is more flexible and in the process of realizing its potentiality.

By recognizing these differences, one can appreciate how different types of matter behave in different ways, and how their capacities to receive forms influence their stability and change.

[11]In Chapter XI, Saint Thomas refers to the Neoplatonists. They developed ideas about creation and causality that differ from those of the Academician. Neoplatonism, which emerged from the 3rd century A.D., is an interpretation and expansion of Plato's original philosophy. Its main representatives include Plotinus, Porphyry, Iamblichus, and Proclus. Here are some relevant characteristics of Neoplatonism:

God as the One: In Neoplatonism, the supreme principle is the "One" or "The Good," which is the source of all existence. Unlike classical Plato, who did not have such an abstract central figure, Neoplatonism places the One in a position of absolute transcendence and as the first cause.

Emanation: Neoplatonists believe that all realities emanate from the One in a process that does not involve mutation or change, but rather a gradual emanation from the most abstract to the most concrete.

Hierarchy of Beings: There is a hierarchy in the order of realities, where the One is the first cause, and beneath it are the *Nous* (Intellect) and the Universal Soul, which are the principles that organize and give form to the world.

Neoplatonists and Christianity

Non-Christian Neoplatonists: The original Neoplatonists, such as Plotinus and his immediate followers, were not Christians. Their philosophy developed in a pagan context and largely in opposition to emerging Christian doctrines. Their ideas on emanation and the hierarchy of beings influenced the religious and philosophical thought of their time, but they were framed within a religious context different from Christianity.

Christian Neoplatonists: As Christianity became established, some Christian thinkers adopted and adapted elements of Neoplatonism to reconcile pagan philosophy with Christian theology. Christian philosophers such as Saint Augustine of Hippo were influenced by Neoplatonism and integrated it into their thought. For example, Augustine adopted the idea of the One as a way of understanding God and emanation as a way of explaining creation and the relationship between God and the world.

[12]Let us remember that intelligible species are ideas or mental representations arising from the activity of the agent intellect. They are the result of the intellect's apprehension of the essences of beings. In the case of humans, they are formed from *phantasmata*, that is, the images captured by the senses. These intelligible species will then be conveyed to the possible intellect, which will form the concept. In Thomistic doctrine, angels do not form intelligible species since they lack senses; instead, those are directly infused in them by God.

www.ingramcontent.com/pod-product-compliance
Lightning Source LLC
Chambersburg PA
CBHW052155220526
45471CB00004B/1694